JOURNEY THROUGH TIME

JOURNEY THROUGH TIME

PART ONE
THINGS THAT MAKE OUR UNIVERSE

Ivica Hausmeister

Matador
9 Priory Business Park,
Wistow Road, Kibworth Beauchamp,
Leicestershire. LE8 0RX
Tel: 0116 279 2299
Email: books@troubador.co.uk
Web: www.troubador.co.uk/matador
Twitter: @matadorbooks

ISBN 978 1788037 587

British Library Cataloguing in Publication Data.
A catalogue record for this book is available from the British Library.

Typeset in 11pt Minion Pro by Troubador Publishing Ltd, Leicester, UK

Matador is an imprint of Troubador Publishing Ltd

I would like to thank to my daugher, Hana,
who gave me the idea to write this book,
and my wife, Vesna, for her love and support.

Contents

PREFACE

I believe that we are quite privileged to live in this time of human history, a time which has allowed us access to an almost unlimited amount of information on whatever topic we choose to read about.

There is no doubt that such opportunity can help us to broaden our horizons. This certainly increased my knowledge and helped me to get a better understanding of the universe we live in.

Sometimes I ask myself how the world would look if we were able to peep at it, just for a short time, some two or three thousand years ahead. I wonder whether our planet would have changed beyond recognition, and into what level of knowledge mankind would have managed to catapult itself.

At the present time of technological development and achievement in science, we have quite a good understanding about our universe and its origins, as much as about our planets, our surroundings and ourselves.

Currently, the so-called 'Big Bang' theory is widely accepted as the most plausible explanation as to how the universe was born. Based on scientific evidence such as the red shift of galaxies, the predicted existence of microwave background radiation and its later discovery, the 'Big Bang' theory is not only the most plausible explanation for the history of the universe, but probably the only possible order of events since the creation of the universe to date.

The development of nuclear physics and the awareness of the existence of a variety of subatomic partials together with progress in a new field of physics, quantum physics, have given another dimension to the existence of the universe, and have helped us to understand current theories.

I myself am a psychiatrist. I graduated medical school and completed postgraduate training obtaining a specialist qualification in Neurology and Psychiatry in my country of origin, then Yugoslavia, before moving to England around twenty-five years ago. I have always been interested in science. Along with medicine, I have been as fascinated by astronomy as by the history of our Earth. Recently, I was looking to find a book that embraces a history of our universe, history of our solar system and Earth going all the way through to modern civilisation. I was looking for a book, which is a kind of encyclopaedia, not written in alphabetical order but in chronological order since the beginning of everything to nowadays.

It was how I came to the idea of writing this book.

I have therefore decided to write this book for the purpose of embracing the knowledge we have about the origin and nature of our universe, our planet and our own existence.

My idea is to write a book that will have a holistic approach to the knowledge we have about the history of the universe, our solar system, the Earth and mankind. In doing so, I do not have the intention of going into details, nor will I be able to do so, as these fields are not my area of expertise.

What I will try to do is to explain, in the way I understand it, what we know so far in a very simplistic way, to make particular physical or chemical reactions easier to understand.

I am not a mathematician or physicist, and I have always tried to understand these topics without using mathematics. This is possible when a layman, such as myself, wants to obtain a basic understanding.

This book is just about that: getting a rough idea about these topics. My attempt was to explain some physical or chemical reactions in a very simplistic way. More importantly, I have tried to give an example for any theory or rules by which our universe is governed. In other words, to understand a particular topic it was very important to me to have an example, which will prove a particular theory, reactions or natural rules logical to me. I was also led by a presumption that my book will be read by people who can remember very little or almost nothing from the science they studied during their primary or secondary school.

As I am not an expert in these fields, I have asked for help from a physicist and a mathematician, to supervise me and correct me. The last thing I want to do is to give misleading information. To this end, I would like to thank my cousin Barbara Lazukić. Barbara obtained a degree in mathematics and physics at university in Slovenia. Her speciality is astrophysics and as such, she was a particular help to me.

I hope that I have managed to take the reader on a journey through time, providing them with a plausible, simple and understandable explanation for the natural phenomena that are taking place in our universe; in the field of physics, chemistry, biochemistry and biology.

INTRODUCTION

This series of books is composed of four parts, which are placed in chronological order, each part dealing with a particular time in the history of the universe apart from the first part.

In this the first part, I have outlined what the universe is composed of, focusing on matter, fundamental forces, energy, dark matter and dark energy. Describing the structure of matter, I have given an explanation of the structure of the atom as the smallest part of an element with a description of elementary particles as the main building block of all matter of which the universe is made up.

I have tried to explain the four main fundamental forces or four fundamental interactions between matter or particles as it is called in physics. As one of the four fundamental interactions is weak nuclear interaction, referring to radioactive beta decay as an example, I have decided to provide some information about radioactivity as well. The remaining three fundamental interactions such as gravity, strong nuclear interaction and electromagnetic interaction are outlined in not too much detail, apart from electromagnetic forces where I found it useful to provide a little bit of the history of magnetism and electricity.

I have outlined the simple definition of energy, focusing on radiation or electromagnetic waves as a way of transferring energy or actually being an expression of a type of energy.

I have superficially explained current evidence supporting the existence of a dark matter and dark energy.

The last chapter in the first part of the book, dedicated to space-time, embraces the conservation of momentum, Einstein's theories of relativity and awareness of the shape of the universe.

Following these introductory topics, in the second of the series, I have started the journey through time explaining roughly the Big Bang theory as an origin of the universe in the second part of the book. Within this part is also superficially outlined the origin of stars' galaxies.

The third part of the series consists of the geological history of our planet, outlining events that took place on Earth in chronological order from the birth of our planet to the present day.

The last, fourth part, of the series contains very superficial details of the evolution of life on Earth with chronological details of the evolution of the human species, again not in much detail but only with an intention to give a rough idea of natural events which take place during the process of evolution.

The books in this four-part series are furnished with illustrations which are my own. There are, however, plenty of wonderful illustrations available online and I was so keen to use some of them for my book but did not know how I would get permission to do so. I have decided therefore to do my own illustrations.

B efore we embark upon this journey through time, we need to have a basic understanding of what makes our universe. Firstly, it would be good to define what we mean or what pictures come into our mind when we read or utter the word *universe* from the aspect of astronomy. I was searching for the best way to define it or perhaps say it in a different way than it has been defined in a dictionary. Whatever way I try to say it, to sound more fancy, the main point is that the universe presents wholeness or totality or everything that exists. Among these things which exist in the universe, we can make a hierarchy and say that the universe is composed of a space and time in which are incorporated matter, dark matter, energy and dark energy, planets, solar system, galaxies and intergalactic space. We are aware that current space of the universe is expanding and that this expansion is accelerated. Thanks to the universe expansion, an idea was born about the origin of the universe from an initial super hot and super dense place, which was less than the size of an atom. This well-known Big Bang theory has now been accepted as the most plausible explanation of history and origin. I will go into more details regarding this scientific model of the history of the universe in Part 2 of this book. Here, I would like to concentrate on those things which our universe is made of.

We can state that our universe is made or consists of the following components:

- Matter
- Four fundamental forces
- Energy
- Dark matter
- Dark energy
- Space-time

It would be natural to talk about space and time or space-time first, but as this is a complicated topic involving Einstein's theory of relativity, I will leave it as the last to be explained in the most simplistic way possible. In this part I have used mathematics, but I have attempted to show it from the angle of a person who has the minimum level of mathematical knowledge.

Before going into each single component, just to outline that calculations made by scientists about the percentage of each of the main components present in the universe show that the universe is composed of around 4% of matter, around 23% of dark matter and around 73% of dark energy.

Although this looks unbelievable at first sight, it is understandable when we think of a dark energy as a space between celestial bodies and space between galaxies. Dark energy is, in fact, referred to as a vast space between galaxies and celestial bodies.

The closest star to our planet is, for example, Proxima in the Centauri constellation (seen on the southern hemisphere). It is 4.3 light years away from us. It means that what we see now is how the star looked 4.3 years ago when light left this star. If we want to know how it looks now, we will need to wait another 4.3 years for the light from the star to reach us. The light year is an astronomical measure or tool to measure distances in the universe. It refers to the distance light will pass during one year of travel. Light travels at a constant speed of around 300 000 km a second. If we multiply this speed by 60 seconds to get 1 minute and then by 60 to get 1 hour and then by 24 (number of hours in a day) and then by 365 (number of days in a year) we will get how many kilometres the light will pass in one year. It is actually slightly less than 10 trillion kilometres or around

9460,730,472,580,800 metres.

Our solar system is located in our galaxy which is called the Milky Way. The Sun is located 28 000 light years from the centre of the galaxy. The closest next galaxy to our own is a small galaxy known as the Large Magellanic Cloud. It is 170 000 light years away from us. In this space there is no object before we reach this galaxy. Galaxies are arranged into clusters due to the gravitational force they have on each other. Our galaxy belongs to a cluster known as the local group. Within this group the Andromeda galaxy, as the largest galaxy, is a little bit more than 2 million light years away fron the Milky Way. The Virgo cluster, the closest cluster to our local group cluster, is 50 million light years away from our own cluster. In this space of 50 million light years there is no detectable object or matter but vast space. Looking at the universe in that way or having that picture in our minds, it is understandable why the universe is composed of 73% dark energy.

1
MATTER

We already have an idea what matter is as we can perceive it with our five senses. We can see things around us; we can touch them, taste them, smell them or hear them. A rough definition of matter is applied to everything which has its own mass and occupies a particular space in a particular point of time. If we leave an ashtray on a table, it will occupy this particular space of a table. This ashtray can be made of glass but also of metal such as gold. While it is on this table we will see it as matter or we can detect its presence at this place by touching it if our eyes are closed. Once the ashtray is removed from this place, then this place will be occupied with a different matter, which makes air. We know that air is composed of oxygen which we breathe in, and other elements such as nitrogen.

An ashtray can also be made of different elements such as gold, iron or different compounds, such as glass.

We have introduced words such as elements and compounds.

In our universe there are many different sorts of matter. They differ between each other as each one has a particular physical or chemical property.

Some matter exists in a solid form such as metal, some in liquid form and some in the form of a gas such as oxygen or nitrogen. Some particular matter is very active chemically and reacts with other matter, creating compounds, while others do not react at all.

The matter that has the specific physical and chemical characteristic is called an element.

In our universe and on our planet have been detected many different elements (more than 100).

The same elements can chemically react with each other, creating compounds. These compounds now have a different property, physical or chemical, from the elements of which they are made.

Hydrogen and oxygen on their own are matter, which comes in gas state. Once they unite in a compound of water, the property of this matter changes; it now comes as a liquid or water.

ATOM

An atom is the smallest bit of matter or element that still has the property of that matter or element. An atom of gold is still gold; an atom of oxygen or helium is still oxygen or helium. Once the structure of an atom is destroyed, as in a process called **fission**, then this particular matter, element, ceases to exist. In fact, in the case of fission, a few new atoms are created that are lighter than the original, and form the smallest bits of other elements.

Fission takes place among a few very heavy atoms. The few new lighter atoms, created in this process, belong to a different matter, or an element with different chemical properties.

We know now very well that an atom is composed of subatomic particles: protons (positively charged particles), neutrons (neutral, with a slightly larger mass than protons) and electrons (negatively charged particles with the smallest mass).

Protons and neutrons make the middle of the atom, or 'nucleus' of the atom, while the electrons go around an atom's nucleus, similarly to how planets go around the Sun in our solar system (Picture 1.01).

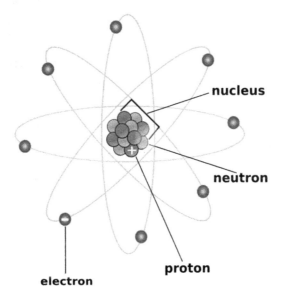

Picture 1.01

Ernest Rutherford, an English scientist and physicist born in New Zealand, contributed significantly to our understanding of the structure of an atom. He conducted an experiment in which he fired alpha rays (which consist of helium nuclei or two protons and two neutrons) at a gold foil. In this experiment, the majority of the radioactive particles went through, some with higher or smaller deflection depending on how close they were to the atomic nucleus. This led to the conclusion that there is a big empty space between electrons orbiting around an atom and an atomic nucleus being in the centre of an atom. Such a big space allowed the majority of positively charged partials to go through an atom with no interference. How big or small deflection is depends on how close or further away to an atomic nucleus these particles are while passing through an empty space of an atom. A small amount of positively charged partials (alpha rays) deflected right back, not passing through the gold foil. These particles hit an atomic nucleus directly and as a result they were bounced back as they were not able to pass through the atoms or gold foil (Picture 1.02).

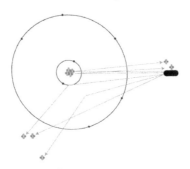

Picture 1.02

It was Niels Bohr, a Danish physicist, who contributed to our better understanding of the structure of an atom and the behaviour of electrons around an atomic nucleus.

Bohr found out that electron orbits are fixed and contain a very precise amount of energy. Electrons can orbit around an atomic nucleus only in these fixed orbits and as long as they are within these orbits they do not radiate energy. If, however, electrons jump from one orbit to the other then they will absorb or emit energy if they jump to an outer orbit or inner one respectively. This energy is expressed as a whole number or quantum.

This single quantum is called a photon. Photons can be described by a mathematical formula, which was created thanks to German theoretical physicist Max Planck. I would, however, avoid using mathematics and try to describe photons in a different way.

I imagine a photon as a quantum of energy which comes in the shape of a package. It is built up purely, or contains only electromagnetic waves, which represent the energy. How much energy a particular photon carries depends on the frequency of electromagnetic waves which make a particular photon. The more waves are in one photon, the higher frequency of these waves is in this particular photon and therefore the more energy this photon carries (Picture 1.03).

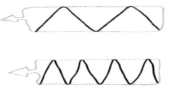

Picture 1.03

9

Wavelength is inversely related to wave frequency. It means the shorter the wavelength, the higher the frequency of waves or the longer the wavelength, the lower the frequency of waves.

We can return now to the main structure of the atom. An atom has positively charged protons and neutral neutron particles, both of which are placed in the centre of an atom, and form the atomic nucleus.

Around an atomic nucleus, but at a greater distance, are arranged negatively charged electrons. They are arranged in 'shells'. Each 'shell' has its own energetic level. The energy of this level increases, going from the inner shells towards the outer shells. The first shell, which is closest to the atomic nucleus, is at a lower energy level than the second shell, which is outside the first shell and further from the atomic nucleus.

Why does each shell gain a higher energy level the further it is from the atomic nucleus?

The way I make sense of it lies in the existence of the positively charged nucleus (protons) and negatively charged electrons. Positively charged protons attract electrons to themselves. So, for electrons to be further away from the positive charge of the protons requires energy, in order for them to oppose the force of the positively charged protons. That is why if they are placed in further shells from the atomic nucleus, these shells have a higher energy level. They need to have more energy to oppose the attractive force of protons.

Electrons absorb energy when they jump from a lower energy level to a higher energy level, i.e. from an inner shell to an outer shell. The reverse happens when an electron goes back from an outer level or shell to a lower energy level or inner shell. In the latter case, an electron radiates the energy that it has previously absorbed to maintain itself at an outer shell. This energy is freed as a photon, with a frequency that is within the spectrum of visible light (Picture 1.04).

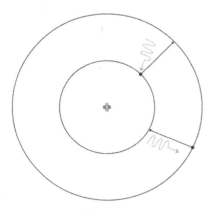

Picture 1.04

An example of this happening in the world is the Aurora Lights, which happen around the Northern or Southern Pole.

What actually happens here is that oxygen and nitrogen atoms get 'excited' by the proximity of some positive particles, which pass near these atoms. This happens during solar winds or storms, when a lot of alpha radiation reaches our planet. Alpha radiation is identified as a helium nucleus, containing protons and neutrons. The Earth is protected from such radiation by its own magnetic field. In contact with the magnetic field, the particles are redirected towards the Earth's magnetic poles. Around that area they come into contact with air and atoms. As alpha radiation has positive particles (protons), they cause the outmost electrons in these atoms to jump into an outer shell. The electron gets an input of energy for this jump, in the shape of a photon. The excited state of the atom is not a natural state, and therefore lasts only for a short period, after which the electron goes back to its natural shell, goes back to its so-called ground state. During this, it emits a photon with a wave frequency in the spectrum of light. If this absorbed energy is high, then the wave frequency will be high, towards the blue colour of the spectrum. In the sky, it will be seen as a blue colour. If the energy of the proton is lower, with a lower wave frequency, it will lean towards the red colour of the light spectrum. We will see a red colour in the sky.

Before going back to the structure of the atom, a question can be asked as to why electrons have to go around an atomic nucleus in fixed energetic level so-called shells.

Why cannot they orbit freely around an atomic nucleus?

The answer to this is that a particular energetic level, shell, protects an electron from falling down to an atomic nucleus being attracted to it by positive charge of protons in a nucleus. To explain it better, I will use our planet and gravitational force as an analogy.

We can imagine our world as an atomic nucleus. The gravitational force of our planet is what brings us down to the surface on Earth, if we fall off a nearby cliff, for example.

We can now imagine an eight-storey car park building, which is built around the equator, going all around it making like a ring around Earth. We can imagine cars (electrons) being driven around Earth at any of these storeys with no problem doing so. If a car is driven from one storey to an upper one, it needs to use more energy (photon input). If it is driven to the lower floor, the storey (shell) does not need this energy any more as the gravitational force of Earth is attracting the car. The car can go down to the next level with the engine switched off (talking about the old type of cars).

Now let us imagine that the storeys (shells) of the car park suddenly disappear. Then all these cars located at different storeys of the car park will fall down to Earth.

This is the same as what would happen to electrons around an atomic nucleus if there were no energetic levels or shells where electrons are orbiting around an atomic nucleus. In other words, electrons would get to an atomic nucleus so the structure of a tome will cease to exist and with it the quality and property of what makes an atom or an element.

So for an electron to orbit around an atomic nucleus with no fear, to be eventually pulled to an atomic nucleus, an electron has to be in an energetic level or shell. An electron cannot survive in between energetic shells so its jump from one shell to another is not a purely mechanical action. What I mean is not a situation where en electron just physically jumps from one shell to another as for this journey an electron has to pass a space between two shells in the process of jumping from one shell to another. This is actually not happening and the process of jumping from one shell to another is more precisely described as the disappearance of a particular electron from one shell and its appearance at another shell. This process is called a quantum leap.

Electrons are arranged into shells around an atomic nucleus. Every shell is, however, further divided into subshells or orbitals. Each orbital can contain only two electrons.

The number of subshells within a shell depends on that shell's number. Shell number one (the first shell around an atomic nucleus) has only one subshell or orbit. Shell number two (the second shell around an atomic nucleus) has two subshells – the one closer to the atomic nucleus contains one orbit, the one further from the atomic nucleus contains 3 orbits. I will not go into more detail regarding this.

The **atomic number** refers to the number of protons a particular atom has in its nucleus. It is what determines a specific characteristic of this atom or element where the smelliest part of that element is present with this particular atom. The charge of an atom is equal zero, as the number of protons equals the number of electrons in a particular atom.

Hydrogen has an atomic number 1 consisting of 1 proton in the nucleus and 1 electron orbiting around. Helium has an atomic number 2 with 2 protons in the nucleus and 2 electrons orbiting around the nucleus.

Carbon has an atomic number 6 with 6 electrons orbiting around it while oxygen has 8 protons in the centre with 8 electrons orbiting around the nucleus.

Elements or their atoms are arranged in the Periodic Table in order of increasing atomic numbers.

It was Dimitri Mendaleev, a Russian chemist, who noticed a repeated pattern of chemical properties among elements, which were known at that time, in the mid-nighteenth century. He initially arranged those elements in order of an increasing mass. Later on that was changed according to an increase in the atomic number.

The **mass number** is the sum of the number of protons and neutrons in a nucleus of a particular atom. Usually, it is easy to memorise that number of protons corresponding to the number of neutrons. For example, helium, with an atomic number of 2, has mass number 4 consisting of 2 protons and 2 neutrons.

Oxygen, with an atomic number of 8, has a mass number close to 16, consisting of 8 protons and 8 neutrons in its nucleus. In reality, an element or atom representing a particular element with a particular atomic number comes in different forms, containing the same atomic number but a different number of neutrons.

Isotops are atoms with the same atomic number but a different number of neutrons.

The atomic number determines the property of an atom and the fact that a particular atom is the smallest part of that element. However, an atom with the same atomic number can exist in many different forms where each atom has a different number of neutrons.

Hydrogen has 1 proton with no neutron. Mostly, hydrogen exists as an element with atomic number 1 and mass number 1. However, in a small proportion there is another form of hydrogen atom with atomic number 1 and mass number 2, consisting of 1 proton and 1 neutron. This is isotope of hydrogen, which is called **deuterium.** There is another isotope of hydrogen consisting of 1 proton and 2 electrons, called **tritium.**

IONS

Not only can the number of neutrons vary within the same elements when they came as isotopes of this element, but also the number of electrons can vary.

An element has zero charge or an atom is neutral as the number of protons in the nucleus is equal to the number of electrons in the shells around the atom. However, an atom can capture en electron in the outer shell from another atom or lose an electron to another atom. Such atoms with numbers of protons and electrons that are not equal to each other are called ions.

If in this process an atom gains an electron, it becomes negatively charged as it has one more electron than number of protons. Such an ion is then called **anion.**

If an atom loses an electron then it becomes positively

charged as it has 1 proton more than the number of electrons. Such ions are called **cations** (Picture 1.05).

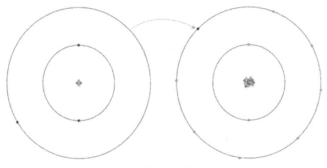

Picture 1.05

What has been described here is chemical bonding, which can be created between 2 atoms. It is called **ionic bonding**. Such bonding is created between metal and nonmetal.

An example of ionic chemical bonding is table salt or sodium chloride.

Sodium has an atomic number 11 with 11 protons in the nucleus and 11 electrons, which are arranged in 2 shells. The first inner shell can have only 2 electrons while the second shell can have up to 8 electrons. With 11 electrons, sodium has 2 electrons in the inner shell, 8 in the second shell and only 1 electron in the third shell. As the third shell is furthest away from the atomic nucleus, it has the less attractive force from the positively charged nucleus. The atom of sodium can therefore easily lose the outside electron and become a cation.

Chlorine has an atomic number 17 with 17 protons and 17 electrons arranged as 2 electrons in the first shell,

8 in the second shell, and 7 in the third shell missing 1, to have 8 electrons in it. With such structure of an atom, less energy is needed to gain 1 electron to get 8 electrons in this shell than to lose 7 to maintain 8 electrons in the outer, which in the case of losing 7 electrons, will be the second shell.

Chlorine, therefore, gains an electron from sodium, becoming an anion, which makes chlorine a strong ionic chemical bond with sodium. Sodium as a positively charged cation reacts with chlorine, negatively charged anion bringing net electric charge to zero. By doing this, we now have molecules of salt composed of these two elements, chemically bound by this ionic bonding.

Ionic chemical bonds compounds when dissolved in water can conduct electricity. If a substance conducts electricity when dissolved in water, then it is called electrolyte.

Atoms can interact among each other by making **covalent bonds**. It happens among those elements or atoms which in outer shells do not have such a small or large number of electrons that they can easily lose or gain electrons respectively. In such cases, atoms enter into chemical reactions where instead of gaining or losing electrons, the atoms share electrons among each other (Picture 1.06).

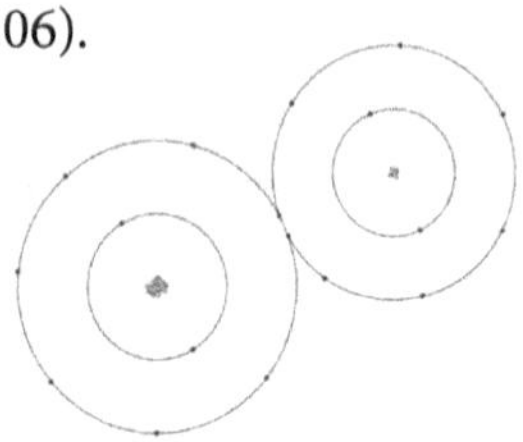

Picture 1.06

An example of covalent bonding

Two hydrogen atoms are bound together by covalent chemical bonds. Instead of gaining or losing an electron, two-hydrogen atoms are in a covalent bond where the creating molecules of two hydrogen atoms share their electrons. In doing so, the outer shells of both hydrogen atoms now have 2 electrons circling around both atoms and in that way establishing the stable state of an element; hydrogen in this case.

We mentioned earlier that the atom is the smallest part of an element which still has the physical and chemical characteristics of that element. The particular chemical characteristic depends on the number of electrons each atom of a particular element has in its own outermost shell.

Apart from the first shell around the atom which has to have 2 electrons to be full, each next shell needs to have 8 electrons in order for that element to be chemically stable, that is to say not to react with other elements.

So-called noble gases such as neon, argon, krypton, xenon, and radon have 8 electrons in their outer shells. Helium is the first noble gas and has only one shells, the first one with 2 electrons. With such atomic structure, they are chemically stable and do not need to get involved in chemical reactions. In fact, it would be extremely difficult to get these elements involved in any chemical reactions.

All the other elements have a number of electrons in their outer shells ranging from 1 to 7 depending on their atomic number and corresponding number of electrons.

All those elements with a number of electrons up to 3 in the outer shell can engage in a chemical reaction where they can lose electrons, forming ionic chemical bonds. They would obviously become cautions in such a reaction.

Atoms can also gain a maximum of up to 3 electrons so all elements whose atoms have 5 electrons in their outer shells can gain 3 electrons, becoming anions and forming ionic chemical bonds.

Carbon with an atomic number of 6 and 6 corresponding electrons (2 in the inner and 4 in the outer shell) is a typical element, which gets involved in chemical reactions by covalent binding.

Many so-called organic compounds, where the carbon atom is the main skeleton or part of a structure of an organic molecule, are made by covalent bonds. Examples are sugar vinegar, carbodioxide (CO_2) or a very complex organic molecule such as DNA.

Ionic compounds are mostly between metals and nonmetals. They are usually solid. The melting point is usually higher than for covalent compounds. Ionic compounds tend to be electrolytes.

Covalent compounds are between nonmetals. They can be solid, liquid or gas. They have a lower melting point and do not tend to be electrolyte.

We now have a rough idea about what is matter and what matter is made of. We have roughly elaborated on different kinds of matter called elements where the smallest part of an element makes an atom.

Atoms can react among themselves, creating compounds. When 2 or more different atoms react chemically among themselves they create a molecule, which is the smallest part of a particular compound having property, which makes this compound unique.

A molecule of water or glucose is still water or glucose until it breaks down to its elements, which initially this molecule was made of. If we go in an upwards direction trying to grasp chemical reactions, different compounds from the simple to the most complicated ones, we will go into the field of chemistry, biochemistry, molecular biology and so on. I do not want to go in this direction and I will go to the core of our matter or downwards to the tiniest particle of which the matter consists. That is within the field of physics of subatomic particles or elementary partials and within the scope of quantum physics or theories.

SUBATOMIC PARTICLES

With modern progress in physics of subatomic particles it is now well known that there are more than 200 subatomic particles. Most of them last only an incredibly short period of time.

Every subatomic particle can be characterised and defined by the three main properties they can express. These are their **mass**, **charge** and **spin** (rotation around their own axis).

With present technology, we are able to calculate the mass of many particles. We know, for example, that the mass of an electron is almost 2000 times smaller than the mass of a proton. We also know that the mass of a neutron is slightly higher than the mass of a proton. They are also massless particles like photons.

Charge is the second characteristic that a particle can express. We are familiar with positively charged particles such as protons and negatively charged electrons. There are also particles which carry no charge and are neutral, such as neutrons.

The mass of a particle is very important, as the quantity of that mass will determine the strength of gravitational force, which exists among particles on a large scale.

The charge of a particle has an important rule when electromagnetic force comes into effect. A neutron, as no charge or neutral particle, has no rule in fundamental reactions among particles, which are carried out by electromagnetic forces.

The third property of a particle, spin, has a very significant importance as it divides subatomic particles into two main groups.

Many particles have their spin, which is quite restricted, and it can come as spin of 1/2, 1 and a 1/2, 2 and a 1/2 integers and so on. So spin can come in a half of a whole number. However, a particle can have a full spin of 1 integer, 2, 3 integers and so on.

Electron spin is only ½ an integer, for example.

All particles that spin ½ an integer, as an electron, or

3 times the spin of electron (which is 1 and a ½ integers) or 5 times the spin of electron (which is 2 and a half integers) belong to categories of particles, which are called **fermions**. Their name is given in honour of Italian physicist **Enrico Fermi**.

Fermion is what all matter in the universe we can see and detect, involving ourselves, is made of.

Fermion is subject to the Pauli exclusion principle, which means that there is a very limited number of fermions that can be placed in a confined space. For example, only 2 electrons can occupy orbitals around an atom.

The first shell around an atom has only 1 orbital and can accommodate only 2 electrons. The second shell has 2 subshells. The first subshell is the so- called S orbital, which can have only 2 electrons. The second subshell within the second shell has 3, so- called P orbitals, each of which can accommodate 2 electrons. With 3 P orbitals in the second subshell and 1 S orbital in the first subshell, the second shell around an atom has altogether 4 orbitals and therefore can have up to 8 electrons around but not more.

The particles can have spin twice that of the spin of an electron, a spin of 1 $(1/2 + 1/2 = 1)$. Such particles with a spin of full integer are called **bosons**. The name was given in honour of Indian mathematician and physicist **Satyendra Bose.**

With such spin, a boson has a property that is not subject to the Pauli exclusion principle. It means that there is no limitation on how many bosons can be accommodated in a confined space. Before elaborating a little bit more on these differences between fermions and bosons, I would like to clarify what we mean by integer.

Integer comes from the Latin word meaning *untouchable*. In mathematics, integer refers to a whole number. The whole number is every number, which is not decimal, or a fraction. It is also 0 or a positive or a negative number, but it has to be a whole number. Numbers 1, 3, 564, or 53952 are integers as they are whole numbers. Minus numbers are integers as long as they are whole numbers such as –1, –3, –35728 and so on.

Subtraction among integer numbers can take place giving an integer number, irrelevant whether it is a + or – number.

For example 10-20 = –10 or 30 –30 = 0. All of these 6 numbers (10, –20, –10, 30, –30, 0) are integer numbers, including 0.

1/2 , 3/2, 5/2, 7/2 are not integer numbers as they are not whole numbers. Obviously, 2/2, 4/2 and 6/2 are integer numbers as they are actually whole numbers (2 divided by 2 gives 1 as a whole number; 4 divided by 2 gives 2 and 6 divided by 2 gives 3 as a whole number).

FERMION has a spin of a half of an integer number which is ½ and forms matter of which all the universe is made, including ourselves. Fermion is subject to the Pauli exclusion principle which is a principle which in the most simplistic way means that two things cannot be in the same place at the same time. As a fermion is subject to such principle, matter which is composed of fermions cannot be in the same place at the same time with another mother.

BOSON has a spin of 1 integer and as such is not subject to the Pauli exclusion principle. Two bosons therefore can be in the same place at the same time. To illustrate the difference between fermions and bosons in that sense, I will use the example of a popular film *Ghost* with Patrick Swayze. He plays Sam, a character who was murdered, but his ghost remained on Earth to resolve the case which was behind his murder. Before Sam died, he was made up of fermion as we all are. As such, he was not able to go through closed doors, or occupied space at the same time as somebody else was sitting there. When he died, his soul remained around as a ghost. Ghosts, for the purpose of understanding this topic, can be described as having a boson property. As boson is not subject to the Pauli exclusion principle, it can go through a closed door because it can be at the same time and place as the other matter.

That boson has such property has been proven in science, in my opinion, by creating superfluid, which was achieved in 1995.

Groups of scientists, including **Eric Allin Cornell, Carl Edwin Wieman** and **Wolgang Ketterle,** were awarded the Nobel Prize in Physics in 2001 for achieving the so-called Bose-Einstein condensation.

In a very simplistic way, the way how I understand this, is that the property of fermions and bosons, or rather to say their behaviour, can be studied at *microscopic* level using the logic of quantum mechanics, or quantum physics.

If, however, it is possible to reduce the temperature of a matter to a temperature very close to absolute zero then the fermion starts behaving as a boson and this property can be experienced at *macroscopic* scale.

I can imagine this or try to understand it in the following way. Electrons, which are fermions, have a spin of ½ of an integer and, as such, only 2 electrons are allowed to occupied each orbit, one with upper ½ of a spin and one with down ½ of a spin. The second shell has 4 orbitals with 2 electrons, which are allowed in each orbital. Therefore,the first shell of an atom can have 2 electrons while the second shell can have up to 8 but no more electrons. If an atom is cooled down close to absolute zero, electrons from upper shells would tend to contract towards the centre of the atom. (The reason I am trying to understand this in that way is that I can understand easily that all things expand at a higher temperature and contract at a lower temperature.) If electrons at a temperature close to absolute zero go towards the centre then they move from the upper shell to the lower shell where ½ of the upper spin of one electron is coupling with ½ of the down spin of another electron, creating the spin of 1 integer, which is the property of a boson.

This is achieved in Bose-Einstein condensation where at such a low temperature particles coalesce into a low quantum state where they become bosons.

This allows the microscopic property of a boson to be experienced at a macroscopic scale.

Eric Allin Cornell, Carl Edwin Wieman and Wolgang Ketterle were able to create a superfluid which has a boson property. Such superfluid looks like it is going through the wall of the vessel (Picture 1.07), like Sam's ghost went through closed doors or occupied the body of Whoopi Goldberg towards the end of the movie.

Picture 1.07

This is not exactly what has happened here. An unlimited number of bosons can occupy the same place at the same time. However, a boson does not occupy the same place a fermion occupies at the same time. If that was the case it would be understandable for superfluid to go through the wall of the vessel. Here, however, we have a different property of superfluid which is the result of reduced viscosity of that fluid to nil and preserved kinetic energy of such fluid. If we stir up superfluid, the created whirlpool will last forever. With such reduced viscosity and preserved kinetic energy, the superfluid tends to climb up the wall of the vessel, come on the other side and go down to the bottom of the vessel until it eventually completely leaves the vessel, dropping out drop by drop from the bottom of the vessel.

ELEMENTARY PARTICLES

Subatomic particles are not the same as elementary particles. Elementary particles are those particles which cannot further subdivide. Before progress in the science of subatomic particles was made there was a time when it was believed that protons were elementary particles which means they cannot be further divided. Today we know that matter is made of only two types of elementary particles or fundamental particles: **quarks and leptons (electrons belong to leptons). Both of them are fermions.**

There are 6 quarks and 6 leptons.

Quarks are: *Up*
Down
Top
Bottom
Charm
Strange

Leptons are: *Electron*
Muon
Tau
Electron neutrino
Muon neutrino
Tau neutrino

There are 6 quarks and 6 leptons, altogether 12 elementary particles from which all ordinary matter is made.

The main difference between quarks and leptons is that quarks are affected by strong nuclear force while leptons or electrons are not.

Quarks unite together to create protons and neutrons. Quarks cannot exist on their own and come as 3 united together, creating neutrons or protons. Neutrons consist of 2 down and 1 up quark. The down quark is slightly heavier which is the reason that neutrons are slightly heavier than protons. Protons consist of 3 quarks as well, but 2 up and 1 down quark.

Quarks also come in pairs, consisting of quark and antiquark. This is a structure of mesons, which are actually bosons, and a carrier of force.

As there are some more names such as hadron and baryon, I would like to outline again that particles can be divided into two main categories depending on whether they are affected by strong nuclear force.

LEPTONS are fundamental particles, which are not affected by strong nuclear force but are affected by weak nuclear force, electromagnetic and gravitational force, the three remaining fundamental forces.

HADRONS are subatomic particles affected by strong nuclear force. They are divided into:

BARYONS, which are fermions, made of 3 quarks such as protons and neutrons.

MESONS are made of a quark and an antiquark. They have a very short life. Their integer is 1 spin and they are bosons.

There are some rules which exist in the world of particles and which cannot be broken. This refers to three conserved quantities. They are: electric charge, Byron number and Lepton number.

1. *Electric Charge* is conserved in a sense that it never changes and always remains zero. An atom has an equal number of protons (positively charged particles) and electrons (negatively charged particles). If ions and cautions are created, they bond together, becoming neutral again with zero charge.

   On a large scale it is gravitational force that has a significant effect in reaction among particles according to the quantity of their masses. Gravitational force has nothing to do with the charge of a particle. If, however, we do not have a zero charge conserved in the universe but different charges among mass in the universe then our universe would look much different to how it is now. We should not forget that electromagnetic force is only 100 times weaker than strong nuclear force and much stronger than gravitational force.

If we are at the top of a high storey building and stretch our hand over the top window with a magnet in our hand with a piece of metal attached to it, we know that this piece of metal will not drop as electromagnetic force is much stronger than gravitational force.

2. *Baryonic numbers* and

3. *Lepton numbers* are also conserved quantity in a sense that the total number of baryons – antibaryons as well as the total number of leptons – antileptons, remains a constant and never changes.

It means that if a baryon is involved in a reaction, then the final result of this reaction is the same number of baryons as at the beginning of the reaction.

An example is a Beta radioactive decay where a neutron is transformed into a proton giving an electron and antineutrino.

N (neutron) = p (proton) + e- (electron) + antineutrino

In this equation we have 1 baryon before the reaction (neutron) and again 1 baryon (proton) after the reaction.

Zero charge is at the beginning of the reaction

(neutron is neutral) and zero charge is after the reaction(it is created + proton and -electron which cancels each other out, giving zero charge).

The baryon or lepton number has to remain uncharged so if we have a reaction which initially did not start with a presence of baryon or lepton, then, even if they appear during the reaction, they have to be in a pair with an antibaryon or antilepton which will result in the cancelling of each other or their annihilation so that the end reaction will not have a baryon or lepton as it was not present at the beginning of the reaction.

In the above example there was only a neutron, baryon at the beginning of the reaction but no lepton present before the reaction. The neutron decayed into a proton and electron (lepton). A lepton is therefore created but also an antineutrino, which is antilepton to electron. Electron(lepton)+antineutrino(antilepton) annihilate each other out with the end result of no lepton present as was the case at the beginning of the reaction.

We know similarly that photons at a particular condition when they collide with each other can create matter. However, every creation of matter in that way has to be accompanied by the creation of antimatter where collision of those two converts them again into energy (photons) from which they were created. The end result will be as at the beginning.

photon + photon = matter + antimatter = photon + photon

Similarly: $0 = 1 - 1$ which $= 0$

Mesons are composed of quarks and antiquarks. They are bosons, as they have a spin of 1 integer. They are carriers of forces for particular fundamental interactions for which they are responsible.

As we mentioned earlier, mesons are made of 2 quarks, actually 1 quark and 1 antiquark. As such, they last an incredibly short time because they annihilate each other. Due to such a short time of their existence they almost cannot be detected. They are called virtual particles. It is accepted theory that particles communicate with each other with the help of virtual particles, which are responsible for carrying and exchanging the information among particles. These virtual particles are therefore the means of communication between particles in a particular fundamental interaction, which takes place between particles. What sort of meson or virtual particle will be involved in an exchange of communication between particles depends on what sort of fundamental interaction between particles is in question. If it is an electromagnetic fundamental interaction, then the virtual particle is a photon. In the case of weak nuclear interaction, the virtual particles are W and Z bosons. With strong nuclear interaction, it is gluon and with gravitational interaction, it is graviton.

The main characteristic of a virtual particle is that its life is very limited and short in duration compared with ordinary particles. It is based on the theory of quantum mechanics and quantum fluctuation in the vacuum.

Quantum fluctuation in the vacuum is referred to as an energy oscillation in a vacuum, which appears to violate a low preservation of energy; also, it is not exactly that preservation of the energy is broken. I have to elaborate a bit on this.

First of all, we have to imagine a container which has been completely emptied in such a way that all atoms are removed from this container. Atoms contain energy. As the container is completely empty it means that it contains zero mass and zero energy. In other words, we have a vacuum where nothing is happening. Well, quantum theory and quantum mechanics state that this statement is wrong and actually so many things are happening in the vacuum. It actually keeps creating matter and antimatter, but as they annihilate immediately the space again remains a vacuum with zero energy, and zero mass.

It looks like something is created from nothing where conservation of energy appears to be violating.

We were describing earlier conservation of charges and baryon and lepton numbers. We demonstrated that if a baryon or lepton number was not present at the beginning of a reaction then the final result of reaction also has to be zero number of baryons or leptons.

If we have created an electron negatively charged and

at the same time a positron, which is positively charged (electron antimatter), then they will annihilate each other, giving zero mass and zero charge. In both cases we have preserved charge and mass as before the creation of an electron there was zero mass and zero charge. In both cases, however, both charges (positive and negative) and masses(matter and antimatter) exist for a limited time. This is the same rule for virtual particles which are created as a particle with a positive energy which is immediately cancelled out and annihilated with antiparticle and negative energy bringing final result to zero mass and zero energy as it was before a virtual particle was created for a very, very, very limited time. In principle, therefore, the law of conservation of energy has not been violated.

Conservation of matter refers to a statement that matter cannot be created or destroyed in chemical or physical reactions. In other words, the amount of matter is not changing in our universe but is preserved.

It does sound very logical. We can look back at the conservation of baryonic and lepton numbers which is very much another way of making the same statement. Baryons (protons and neutrons) together with leptons (electrons) make atoms. An atom itself is a very stable structure. When atoms of hydrogen combine with atoms of oxygen to create a molecule of water, they do not disappear but just become a part of a water molecule. It means that the sum of masses of 2 hydrogens and mass of 1 oxygen (number of those atoms in a molecule of water) is equal to mass of molecule of water.

It was **Antoine-Laurent Lavoisier**, a French scientist and chemist, who revolutionised progress in chemistry, as a discipline within natural science, in the 18th century by discovering this law. He did a very precise measurement of reactants (substances or elements used in chemical reaction before the chemical reaction), end product (result of chemical reactions between reactants). He confirmed that the amount of reactants is equal to the amount of product, meaning that there was no loss or creation of matter. The only difference was that product was molecule of compound with a new physical or chemical property, which has the same amount of matter as sum of matter, which takes part in this reaction. In the above example, 2 hydrogen and 1 oxygen together, were equal in amount to the amount of one molecule of water.

The same applies to physical reaction. If we measure an amount of ice before melting and then the amount of water obtained after the ice is melted, we will find out that it is an equal amount, meaning no matter was lost or created.

The conservation of matter is also called conservation of mass. It means that mass and matter are interchangeable which is understandable as mass is a fundamental property of matter. But what is actually mass? How can it be defined?

The Encyclopedia Britannica offers a very nice definition of mass as a quantitative measure of inertia. What is measured is a resistance that body of matter offers to a change of its speed or position upon the application of a force. It is based on Newton's first law of motion that states that every matter tends to remains in motion at a constant speed or stand still indefinitely unless it is acted upon by force. Law of inertia, Newton's first law of motion.

An example is a parked car and a little child in a toy car next to it. Both are still, not moving. If we want to change their inertia, in other words, move them, then obviously it would be much harder to do it to the car, which has a larger mass, then to the child in the toy car with a much smaller mass. The same applies if we want to stop them or change their inertia of motion.

Definition of mass is very tricky and can be very confusing, particularly following Einstein's theory of special and general relativity.

Basically, if we do not consider Einstein's theory of relativity, we can say that mass can be manifested in two ways. Namely as:

Inertial mass, which can be measured when we apply the force to move body of matter.

Gravitational mass, which is a measure of mass proportional to the gravitational force of attraction between two bodies, making them move towards each other.

Weight is not the same as mass. It refers to a force experienced by the matter due to gravity and it could be interchangeable with gravitational mass.

An example of inertial and gravitational mass could be described in the following example:

Our weight will be 6 times less on the Moon than on Earth as gravitational force is 6 times less there than on Earth (gravitational mass). We will equally need more force on the Moon to move objects, like a bus compared with a motorbike, as is the case on Earth (inertial mass). Both of them will be, however, 6 times lighter on the Moon (gravitational mass).

When we take Einstein's theory of relativity into account then we have to introduce two more concepts, which are:

Rest mass refers to the mass of a matter which is moving at the same speed as an observer.

What does this means?

It means that if my wife measures my weight on a train, which moves at a speed of 100 miles per hour, my weight will be 90 kg for both my wife and myself (and scale which is on the train) as we are moving at a speed of 100 miles per hour. My weight would be the result of gravitational mass.

Relativistic mass refers to a mass which is the sum of rest mass, and mass increased due to speed. In the above example, my mass is significantly increased for a person who is next to the train I am in and passing him by at a speed of 100 km per hour.

The average punch of a professional boxer is 400 kg and serves as another example. The average weight of this boxer can be 90 kg but it is the speed of his fists that adds to the rest mass, making an average mass to be around 400 kg.

Einstein has completely changed our awareness about the relation between mass and energy. This affects the definition of mass and energy conservation and the correctness of such a statement if we look at these laws separately, referring only to mass or energy.

Einstein's famous equation defines so-called **mass energy equivalence**

$$E = mc^2$$

The equation was derived from the special theory of relativity. The equation clearly states that energy is equivalent to the amount of mass. We know that **c** is the speed of light, which is constant and unchangeable. It is also squared, giving a very big number. This means that a very small mass has a very large amount of energy.

We can, for a moment, leave out **c** squared, as it is constant, unchangeable, and look at the relation between energy **E** and mass **m**. In this scenario, if we want to increase energy, we have an increase in mass. If we have less energy, we have less mass.

With mass energy equivalence derived from the special theory of relativity, we cannot any more formulate or state that conservation of mass means that mass cannot be created as an increase in energy increases the mass and a reduction in mass releases and reduces the energy, which was locked away in the mass. We therefore talk about **mass-energy conservation**.

With the development of nuclear physics we are able to measure very precisely the mass of reactants taking place in a chemical reaction and the mass of the product of this reaction. It is established that the sum of masses of reactants is slightly larger than the mass of newly created compound, product. The reason for this is a need for energy to be released from reactants, which will bind them together into a new molecule. The only place from where this energy can come is the mass of reactants that are involved in this particular chemical reaction. So when the binding energy is unlocked (released) from some of the mass of reactant, their masses are subsequently decreased, giving the mass of product to be slightly less than the sum of masses of reactant before the chemical reaction.

If we put this another way, we have to bring energy to separate atoms or reactants from the product of new molecules created in a chemical reaction.

Molecule of water + binding energy = hydrogens + oxygen

We need to do something to a molecule of water to separate atoms from the molecule. We need to bring energy back, which was released to bind them together in a molecule of water. By doing this we will succeed in reversing the reaction with the disappearance of binding between these atoms and the reversal of these brought energy to the mass of oxygen and hydrogen. Now separated, atoms of hydrogen and oxygen will have their masses again at the amount it was before they got involved in the reaction.

In this example the mass-energy is conserved but during the chemical reaction there was change or reduction of mass and its increase in the reverse process with subsequent reduction or release of energy from the mass and its increase and absorption into the mass respectively.

I will touch upon this topic once more at the end of the third chapter.

2
FOUR FUNDAMENTAL FORCES

Four fundamental forces in the universe are:

Strong nuclear force
Weak unclear force
Electromagnetic force
Gravitational force

Fundamental forces can be defined as fundamental interactions among particles of matter. These fundamental interactions are maintained by exchange of carrier between particles.

We have division of matter into two main categories: fermions and bosons. **Fermions have a job to make up our matter while the boson's job is to carry force and energy.**

The fundamental interaction between fermions or matter, which binds them together, can be compared to football players, for example. The players are fermions who are bound together during play by throwing and catching a ball. The ball carries force and energy from player to player. Therefore, the ball is a boson. The range of a particular fundamental force is determined by the mass of the boson which carries this particular force. The bigger mass the narrow range of this particular force or lighter the boson which carries this particular force, the longer range of this force.

The boson which carries an electromagnetic force is a **photon** which is massless. The boson which carries gravitational force, **graviton**, is also massless. Both gravitational and electromagnetic forces are therefore unlimited.

The weak nuclear interaction of weak forces is referred to as radioactivity of elements. Elements with a high nuclear mass such as Uranium's isotops can be very unstable which makes them undergo spontaneous nuclear decay. During this process we can have radiation in the shape of beta emission or so-called beta decay. Weak nuclear interaction is caused by **W** (W stands for weak nuclear force) and **Z** bosons. They are among heavier elementary particles, their mass around 100 times heavier than a proton. With such a mass of W and Z bosons as a carrier of weak nuclear forces, the range of nuclear forces is low.

Strong nuclear forces or interactions have a massive carrier within the nucleus called **the pion.** The strong interaction has an unusual behaviour to grow in strength as the distance increases between particles or quarks. **The gluon** is a force carrier with the responsibility to hold quarks together. All the other three remaining forces are known to decrease in their strength with the distance but strong nuclear force does not. That could be understood if gluon, their carrier which attracts quarks together , is imagined as an elastic band which is placed among quarks in that way that one elastic band is attached between two quarks. When quarks are tightly located to each other, three of them, making proton or neutron, the elastic band (gluon, carrier of strong nuclear force) is relaxed and their existence is almost not noticable. If, however, three quarks want to separate, then the elastic band comes into force, getting tight, pulling quarks back together. Although the strong nuclear force is stronger with the distance, this refers only to small distances within the atomic nucleus.

The strong nuclear force is 100 times stronger than electromagnetic force. This is understandable as they need to keep together protons within the atomic nucleus and overcome electromagnetic force or the repulsion of protons against each other due to the fact they are the same charge.

The strong nuclear force is stronger than weaker nuclear force by factor 10 on power of 11 which is 10 000 000 000 0 times stronger than weak forces. However, a strong nuclear force is 10 on power of 41 stronger than gravitational force.

If we use the whole number, then strong force is stronger than gravitational force 10 000 000 000 000 000 000 000 000 000 000 000 000 000 0 times.

If we outline four fundamental forces in order from the weakest one to the strongest one then they can be written as follows:

Gravitational force < Weak nuclear force
< Electromagnetic force < Strong nuclear force

I will outline these forces in order from the weakest to the strongest one.

GRAVITATIONAL FORCE

This is the weakest force but at the same time the long distance force. It is interaction between two bodies which have masses . Due to the fact that these bodies or matter have a mass , the gravitational force exists among these two bodies.This gravitational force is expressed by an attraction between these two bodies. The strength of this attraction depends on the product of these two masses divided by square of distance between these two bodies. If both bodies have a constant mass, then the force of gravitational attraction is inversely proportional to the square of the distance between these two bodies. If an object is positioned 1 km from Earth and a similar object with the same mass 10 km away from the Earth then an attraction of gravitational force to the object at the distance of 1 km will be 100 times stronger than on the object which is positioned 10 km away from the Earth.

Gravitational force is important at the level of the universe as it is the force that shapes the universe. It is thanks to gravitational force that all celestial bodies are created (stars,planets, satellites). It is force that holds the solar system together as much as galaxies.

Isaac Newton was the first scientist who become aware of the existence of gravitational forces. With his work on mechanics and gravitation through precise formulation of his three laws of motion, Sir Isaac Newton revolutionised physics. This opened a new chapter where the universe was to be understood more clearly through existence and the effect gravitational force has on a large scale. However, the final touch in adjusting the theory relating to gravitation to its perfection came from Albert Einstein with his work on the theory of relativity. He realised that space, which has a third dimension, has another, fourth dimention attached to it ,which is time. Time, as a fourth dimension, is in relation to the other three through the speed of light which has only an absolute value. Unlike classic physics which postulates that space and time are absolute and not changable, Albert Einstein made famous a mind experiment which led to the conclusion that light does not change its speed and is always 300 000 km per second regardless if it is measured from the spot which is at rest compared to the light coming, or if it is measured from the spot which is moving in relation to the speed of coming light. He also predicted that light, as a matter, is subject to the force of gravitation and as such it can be pulled when it passes near a large celestial object such as a star. In that case, light is bent which can be perceived as a shift of a place where a particular star is usually seen when her light comes to us with no bending. In other words, when light is not passing near a large celestial body or star, then there is no displacement of this star from where the light is coming as there is no bending. (This phenomenon of bending and subsequent shifting of the position of the star is called gravitational lensing, as reminded by the displacement of a background object when we put a lens between our eyes and background objects.) This particular prediction Einstein made was tested by English astronomer Arthur Eddington during an eclipse on 29th May 1919. He went to the island of Principe, off the coast of West Africa, where a total eclipse was expected to take place on 29th May 1919. The main principle of a measurement was to take a photograph of stars which are in the background of the Sun when the Sun is present ,and the pictures of the same stars when the Sun is not present, at night. Obviously, with the presence of the Sun it is only possible to take photographs of stars if the Sun is in a total eclipse, such as took place in this part of the world on 29th May 1919. Following the success in taking pictures of the Hyades constellation before and in the presence of the Sun, the measurement Arthur Eddington made showed that the Sun caused a deflection of 1.61 seconds of arc between the same stars in absence of the Sun and their position on the photographs when the Sun was present. This deflection was close to a deflection of 1.75 seconds of arc as Einstein predicted.

Hyades is a star cluster which is within the constellation of Taurus. It is easily spotted in the sky as a V shape but the V is more in a horizontal position . It hasa giant red star called Aldebaran which does not belong to the Hyades

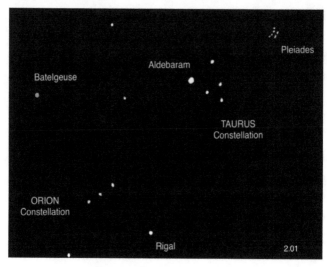

Picture 2.01

cluster but is so large as it is placed much closer to us than the Hyades cluster. The Hyades cluster is otherwise closest to our solar system and is composed of hundreds of stars of the same age

WEAK NUCLEAR FORCE

Weak nuclear force or weak nuclear interaction is caused by the emmision or absorption of virtual particles or W and Z bosons.

Beta radioactive decay is a reaction which takes place as a result of a weak nuclear interaction. In this process the nucleus of an atom is decayed in a way that a neutron is transformed into a proton. During this process an electron is emitted from the nucleus as well as neutrino.

Virtual particles of weak nuclear forces have the ability to change the flavor of quarks. We know that quarks come in up or down flavors and that a neutron is composed of 2 down and 1 up quark while a proton is composed of 2 up and 1 down quark. W bosons can change down in up quark and vice versa. With such property, weak nuclear forces are important in radioactive decay such as beta decay and also in nuclear fusion of hydrogen atom to hellium. (Atom of helliun has 2 protons and 2 neutrons.) Therefore, in order for nuclear fussion of hydrogen atom to take place, it needs to be created deutherium atom or hydrogen isotop with neutron as an additional particle so that the final result of the fusion will be hellium with 2 neutrons and 2 protons in atomic nucleus. To have deiterium we need weak nuclear interaction to take place or bosons of weak nuclear interaction to change up quark in down one and in doing so transform 1 proton into a neutron.

Weak nuclear interaction has an important role in beta radioactive decay as well as in nuclear fusion (explained above) which takes place in a star. The nuclear fusion is the important fuel of a star, preventing the star from collapsing due to the powerful gravitational force of an enormous mass of which a star is usually made. As weak nuclear force plays an important role in radioactive decay or a type of radioactvity (beta decay), it is the right place to elaborate a bit on radioactivity.

RADIOACTIVITY

Radioactivity is a property of some matter to emmit subatomic particles or energy spontaneously.

It happens among those elements which have a high atomic mass or a large number of protons and neutrons in its nucleus. When the number of protons and neutrons increases above binding power which keeps them together within the nucleus, then this nucleus becomes unstable and will decompensate or decay into a more stable configuration.

The binding power which keeps protons and neutrons together in the nucleus is a remnant of the strong nuclear forces which keep quarks together in a proton or neutron. It could be described as a spillover of that power beyond boundaries of protons or neutrons, having 100 times stronger power than electromagnetic forces. As such, it easily overcomes the power of electromagnetic forces which tend to pull protons apart from each other. However, this strong nuclear interaction is powerful only for a short distance within the nucleus in a range or diameter which is not larger than the sum of 2 and a half protons diameter (Picture 2.02). As more protons and neutrons unite in an atom of heavier elements, then their closiness reduces as they begin tightly to pack in the nucleus. As the tight connection between protons in the nucleus enlarges or the distance between protons decreases , it takes about sixteen nucleons (protons and neutrons)to be in the nucleus of an atom to reach the size of 2 and a half of the proton's diameter. When the

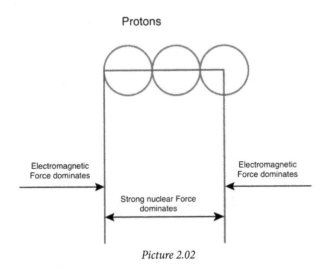

Protons

Electromagnetic
Force dominates

Electromagnetic
Force dominates

Strong nuclear Force
dominates

Picture 2.02

diameter of the atomic nucleus reaches that size, then strong nuclear forces start to lose their strength and an electromagnetic force takes dominance. The last element in the periodic table (from the smallest to the higher atomic number) which is still stable is iron as strong nuclear forces and electromagnetic forces are in a kind of eqilibrium or have an equal effect so the nucleus of iron is still stable. However, it is not before an atom reaches a mass of around 207 that it becomes so unstable that it starts spontaneously to decay. When this process starts, it takes the form of emitting particles (alpha, beta particle), energy (gamma rays) or splitting into 2 so-called daughters or 2 atoms with stable atom mass which is smaller than it was in the mother atom or initial atom, the process called **fission**.

Alpha radioactive decay is the spontaneous decay of an atom where alpha rays or particles are emitted. Alpha particles are the nucleus of a helium atom, having an atomic mass of 4 and atomic number 2 (they consist of 2 neutrons and 2 protons). During such a process, the initial element loses its mass by 4 and an atomic number by 2. The final result will be a new element with reduced atomic number by 2 and mass by 4.

Uranium with the atomic number of 92 and mass of 238 decays, emitting alpha particle or helium nucleus of atomic number 2 and mass 4 and gives another element thorium with atomic number 90 and atomic mass 234.

$$^{238}_{92}U = {}^{4}_{2}He + {}^{234}_{90}Th$$

In this equation of alpha decay of uranium to thorium we can demonstrate again the conservation of the baryon number. The mass and atomic number of uranium is equal to the sum of the atomic number and masses of helium and thorium.

Unlike beta decay, alpha radioactive decay is not the result of weak nuclear interaction where W or Z virtual particles play a role. Instead, it is caused by so- called quantum tunnelling.

Quantum tunnelling refers to a particular property of subatomic particles where the particles can go through a barrier with no additional energy needed for this to happen.

We can use the analogy of playing with a tennis ball which we are repeatedly throwing against the closed door. In each case, the tennis ball will bounce back from the closed door. It will not go through the door. In order to do so, the ball needs to have much more energy. If we put the ball in a gun from where the tennis ball can be fired at the speed of a bullet, then the ball can pass through the closed door making a hole in it. In other words, it does have enough energy to pass through this barrier.

In a world of subatomic particles where distances and particles are fantastically small, different rules are applied. The movement and the position of subatomic particles do not follow the rule of classic mechanics, which is applicable at great distances in the world we know. They follow the rule of quantum mechanics.

What does that mean?

Well, we can say that unlike in our world where we can easily define the position and the momentum of an object, such as a tennis ball (meaning the ball's location and its movement), this is not possible in a micro world where subatomic particles rules the world. Electrons, protons or atoms can be in any position and have any movement. There is, therefore, a great uncertainty as to where any particular particle is located, or rather to say, it could be located anywhere. This, however, implies the possibility that a particle can also be found on the other side of the barrier, which means that the particle has to pass the barrier with no use of any extra energy. This is actually quantum tunnelling.

With such unlimited possibilities of the particle's location or rather the uncertainty of its location, quantum mechanics helps to calculate the probability that a particular particle will be at a certain place, including the probability that it will be on the other side of the barrier; in other words, the probability that quantum tunnelling will take place.

It was **Richard Feynman**, an American theoretical physicist, who developed the part integral formulation and an equation, which is helpful to determine the probability that a particular subatomic particle will be at a certain position.

Professor Brian Cox, a physicist, has outlined a simple version of Richard Feynman's equation in a TV documentary *A Night with the Stars*. That was regarded as a full lecture delivered to a celebrity audience in Manchester and was aired on the BBC. The programme can be easily found on YouTube and I would strongly recommend it.

In this programme Prof. Brian Cox outlined a simple version of Feynman's path to integral formulation. He wrote a formula of a time needed for quantum tunnelling to take place or the probability that after a certain time the particle or object can be on the other side of the barrier. According to an outlined equation, the time needed for this to happen depends on the product of the distance we expect the object to jump, the size of the place where this object is located and the mass of this object divided by the Planck constant.

$$T > \frac{x\ dx\ m}{h}$$

x is distance to jump, dx is the size of the location where the object is positioned and m is the mass of an object while h is the Planck constant which is 6.6 x 10 on power of −34.

As all matter is composed of atoms and atoms follow the rule of quantum mechanics, then every object, does not matter how large or massive it is, can have a probability to be located on the other side of a barrier at one point in time. The time needed for this possibility to happen depends on the size of a location in which an object is placed, the distance it needs to jump to be on the other side of the barrier and its own mass.

In the case of a tennis ball to be found on the other side of the closed door, we calculate the size of the room, say it is 4 metres, and multiply this with the distance we want the ball to be in order to be outside the closed door. That could be 4 metres and 20 cm, for instance. We then multiply all of this with the mass of a ball, say 10 g, and then we divide this with a Planck constant. The result obtained is the time needed for the ball once to be found outside the closed door during our time of throwing the ball on the wall. I have no intention of making this calculation here. Instead, I will use an example Prof. Brian Cox gave in the documentary programme mentioned. He used the example of a diamond being in a box of 5 cm, wanting to calculate the time needed for the diamond to be found 6 cm away (meaning outside the box) with the mass of diamond being 60 grams. The product of these values divided by the Planck constant has given the result of 3 x 10 on power of 29 seconds which meant that this number of seconds was needed for the diamond to be found 6 cm away or outside the box. Prof. Brian Cox outlined that this number of seconds was equal to a 600 billion times the current age of the universe. Such a long time is due to such a small number of the Planck constant. If, however, the size of location of the object, together with mass of the object and size of the distance needed for the object to jump is incredibly small as in the case of subatomic particles, then the time for the particle to be on the other side or for quantum tunnelling to take place is not so long. It happens therefore more frequently.

Quantum tunnelling is therefore what facilitates alpha radioactive decay.

Quantum tunnelling also facilitates the nuclear fusion process as much as weak fundamental interaction.

The fusion of hydrogen nucleus into helium nucleus takes place in the centre of a star. In a star's core there is very high pressure and temperature, which gives high energy input to hydrogen atoms and electrons rotating around its nucleus, which consists of one proton. This high energy, thanks to the high pressure and temperature, makes an electron jump from a lower energy level to the higher energy level of a hydrogen atom (it jumps from a shell closer to the atom to a shell further from the atom). As the temperature is very high and therefore energy, an electron moves in cascade from the first to the last shell, the 8th shell. Eventually, with constantly high energy, en electron jumps from the last shell decupling itself from the hydrogen atom. The hydrogen atom has lost its structure and we have now plasma formation consisting of protons and electrons. Also, protons now move rapidly in all directions; they are prevented from coming close to each other due to electromagnetic force and consequent repulsion among each other. The barrier of this repulsion is the distance among two protons equivalent to the sum of radius of 2 and a half protons. It is thanks to quantum tunnelling that protons find themselves on the other side of the barrier or at a distance less than 2 and a half of radius of proton (Pic 2.02). Once they are within this distance they are captured by strong nuclear fundamental interaction, which are strong within this radius. That is how and why Quantum tunnelling is important for nuclear fusion to take place.

Beta radioactive decay takes place when weak fundamental interactions are involved. Depending whether W and Z bosons change down to up quarks or up to down quarks, we have Beta-minus decay of Beta-plus decay respectively.

Beta-minus radioactive decay is a process where the neutron of an atom is transformed into a proton and emits negatively charged electrons and antineutrino. Bosons of weak fundamental force or mesons change the flavor of down quarks to up quarks, transforming neutrons (2 down+1up quark) into protons (1 down+2 up quark). Electron and antineutrino is emitted. Here, the initial atom has the same mass as the daughter atom which is the end product of this decay but increased atomic number by 1. An example is beta-minus decay thorium with mass of 234 and atomic number 90 to protactinium with mass of 234 and number of 91.

$$^{234}_{90}\text{Th} = {}^{234}_{91}\text{Pa} + \bar{e} + \bar{v}$$

$\bar{e}$ is electron while $\bar{v}$ is antineutrino

Beta-plus radioactive decay or **positron emission** does not occur among naturally present isotopes but does take place among artificial ones made by humans. Here, mesons of weak fundamental interaction, virtual particles or bosons change up quark to down 1, transforming protons into neutrons. The final product is an element with the same mass but reduced atomic number by 1. It emits positron (positively charged electron) and neutrino. The beta particle in this case is positron. It is a case of artificially made isotope of kalium (potassium) and its decay by beta-plus radiation to argon with the same mass and reduced atomic number from 19 to 18 (by 1).

$$^{40}_{19}\text{K} = {}^{40+}_{18}\text{Ar} + e + v \text{ (neutrino)}$$

Electronic capture is a similar process to Beta-plus radioactive decay. The difference is that there is no positron emission and that an electron from the cloud or orbital next to the atomic nucleus is captured and fused with a proton in the nucleus, giving a neutron. As the orbital from where an electron is captured is now empty, the atom is in an excited state. An electron from the other orbital takes the place of this electron, which is captured, bringing down an atom in a ground state. In this process, x-ray is emitted.

Gamma decay is radiation, which can be emitted following alpha, or beta decay or it can happen on its own. It is the result of newly produced atoms being in an excited state and its tendency to go back to ground state. During this process, energy is emitted in the shape of gamma rays. Atomic nuclei can be in an excited state on their own. In such a case, when they return to ground state they emit gamma rays. High energy level within the nucleus itself is not understood well as high energy levels of electrons. As we know, high energy levels of electrons are referred to as quantum energy levels which are described as shells around an atomic nucleus. In a similar way, an atomic nucleus has a different quantum energy level where protons can go on a higher energy level of being excited. In such a state, protons tend to go back to a low energy level or ground state, as this is a natural law in the universe. In this process of going back to ground state, they emit gamma rays.

Spontaneous fission is a type of radioactivity where the nucleus of a heavy element splits into two roughly equal newly created nuclei, each of which is around half of the mass of an initial heavy atom. This process happens very rarely among natural heavy elements such as uranium but does happen among artificially created heavy elements such as fermium-256. In the process there are always a few neutrons that are left out. Due to the acceleration they got in the process of splitting of an initial atom, they hit other atoms, causing them to become unstable and split. This acceleration is due to the strong repulsion of split atomic nuclei due to their same charges and the fact that 2 newly formed nuclei are now at a distance more than 2 and a half radius of proton where strong repulsive forces of electromagnetic force dominate. As neutrons accelerate as well they can hit and reach the inside of new nucleons causing them to split by emitting new free neutrons which will hit other atoms causing them to split. That is a chain reaction.

Otto Hahn, German chemist, and **Fritz Strassmann**, German physical chemist, discovered nuclear fission in 1938 around Christmas time. Following this discovery and further progress in nuclear physics, an atomic bomb was created. On 16th July 1945, the first atomic bomb was tested near Alamogordo in southern New Mexico.

On 6th August 1945 the atomic bomb was used as a weapon for the first time. It was dropped on Hiroshima, instantly killing around 70 000 people with the number rising above 100 000 by the end of the year. The second bomb was dropped two days later on 9th August 1945 on

Nagasaki. It killed around 40 000 people. It marked the end of the Second World War.

ELECTROMAGNETIC FUNDAMENTAL INTERACTION

After strong nuclear force, electromagnetic nuclear force is the most powerful force among the three remaining ones, including electromagnetic force itself.

Unlike other forces, it refers to the interaction between charged particles, protons and electrons, and does not affect neutral particles, neutrons. This fundamental interaction takes place with the help of virtual particles (bosons or photons or light particles) which exchange communication between particles telling them to attract to each other or repulse depending whether they are the same or opposite charge. When two electrons are on the path of collision, a virtual practical photon is created. Photons inform particles (electrons) that they should repulse each other as in the diagram known as: Richard Feynman's diagram.

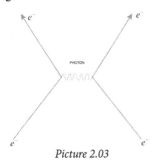

Picture 2.03

An electromagnetic force keeps the structure of the atom together as well as that of complex structures such as molecules. By doing this, it helps matter to appear in different shapes and forms including leaving world and us.

The strength of the forces is, as with gravitational forces, inversely proportional to the squared distance between particles affected by the force.

Looking from the aspect of classical physics, electromagnetic force is the combined force of two phenomena, which occur in the universe: electricity and magnetism. Perhaps the best way to describe electromagnetic force in relation to electricity and magnetism is if we refer to the electromagnetism as a coin with two sides: one electricity and one magnetisms.

Both electricity and magnetisms were known well before it was realised that these two phenomena are interconnected. It was not until the 19th century that these two phenomena were seen as united forces of electromagnetisms. There were three crucial historic events, which help us understand the connection:

1. Accidental discovery confirming the relationship between electricity and magnetism. **Hans Christian Orsted,** a Danish physicist and chemist, noticed that a compass needle deflected from magnetic north when it was closed to electric current. This discovery was made during winter in 1819-20. He is regarded as the first scientist to notice this connection. However, it was an Italian scientist and philosopher, **Gian**

Domenico Romagnosi, who first noticed a magnetic effect of a current in 1802. His work was published in a newspaper, which was not so popular, and his work was overlooked by scientific society at that time.

2. **Michael Faraday** discovered that he could produce electricity from changing the magnetic field. He performed this experiment of inducing current by changing the magnetic field in 1831.

3. **James Clark Maxwell** gave the final touch to comprehension and realisation of the existence of combined electromagnetic force. Through experiments and his equations, Maxwell realised that a change of electric field produces a change of magnetic field, which in turn causes a change in electric field and vice versa. He predicted the existence of electromagnetic waves, which propagate through space at the speed of light. It was finely understood the nature of light. The light was nothing more or less than electromagnetic waves with a frequency of waves or their wavelength within the spectrum of visible light.

As electricity and magnetism were known well before electromagnetism was discovered, I would like to pay brief attention to each of them with a brief history of the events which shape our knowledge and understanding of these phenomena through the centuries.

Electricity is a result of the changes, created in a system and tendency to keep a zero charge within the system.

We have discussed conserved quantity in Chapter 1. We mentioned under the heading of subatomic particles that one of conserved quantity is zero charge as there is an equal number of protons and electrons in a closed system (a metal bar, as an example of a closed system) or universe, which is a closed system on a maximum scale.

The distribution of negative and positive charges within the system can be changed with one side becoming more positively charged and the other becoming more negatively charged. In such a situation there is a tendency to bring charges to zero within the system, which will create movement of positively charged particles and negatively charged, one towards each other. As positively charged particles are protons, which are trapped in a nucleus, it will be very difficult for them to move. Electrons, on the other side, particularly ones on the outermost shell around an atom, can move freely. They are therefore the ones which move along the system towards positively charged particles with a final go to create zero charge within a system. The movement of electrons within a system from negative charge to positive charge is called **Electric current:** a phenomenon that takes place when we have electricity. We can have positive and negative charges within the system and they can remain as such with no possibility to go back to zero charges. This happens in situations where between these charges is a material composed of such atoms or molecules, where electrons cannot move freely. Examples are rubber or plastic, which surround electric wires in sockets. Such material is called an insulator.

Materials which allow free movement of electrons are called conductors. Metals such as copper and iron are good conductors of electricity.

When we discussed type of bonding between atoms to create molecules in Chapter 1, we mentioned two main types of bonding: ionic bonding and covalent bonding. This bonding takes place thanks to the tendency of the outermost shell of an atom to have a maximum number of electrons (8 for every next shell apart from the first shell which is 2). In this process only, therefore, electrons on the outermost shell take part in bonding. These electrons are called the **valence** electrons.

In addition to ionic bonds, specific to metal reacting with nonmetal, and covalent bonds, specific to nonmetal bonds, there is a bond within metal atoms similar to covalent bond, called metallic bonding.

Unlike typical covalent bonding where electrons are shared among two or more atoms within a molecule, in metallic bonding, each atom of metal donates its valence electrons to an electron pool, called a sea of electron, which can freely move across whole material or systems (metallic bar, or iron bar). This is a reason that metals are a good conductor of electricity.

Static electricity is created from excess charge. Within the system we do have an equal amount of positive and negative charges and that cannot be changed. What we can do, however, is to move these charges around and, by doing so, create imbalance in charges in a particular system. As we stated earlier, we cannot move positive charges as they are part of the atomic nucleus, but we can move electrons.

If we use a glass rod and rub it with silk for a while, then a number of electrons from the silk will jump on the glass rod. The rod will become negatively charged, as it will have a surplus of negative charges.

We said that within the system, charge is zero. The glass rod is a closed system. However, charge can be changed in a system if it is not closed and influenced by another system from outside (silk). The glass rod now tends to regain zero charges and in contact with a ping-pong ball will pass these electrons on the ball, making it become negatively charged.

We have the opportunity to experience static electricity in everyday life. On a day after combing hair it can happen that our hair goes up with separation of hair. This happens as a number of electrons pass from comb to hair. As electrons repulse each other, each hair tends to go away from the others. As soon as we touch a material where surplus of electrons from our hair can be passed to, we bring charge of hair to zero, making hair going down take the position it had before becoming charged.

We have already mentioned electric current which is created when electrons move through electric wire.

Historically, although there is some evidence suggesting that people were familiar with static electricity, there was not a significant curiosity for these phenomena before the 17th century. **Thales of Miletus** has experimented with static electricity in about the 6th century BC, but first experiments with electricity started in the 17th century. **Otto van Guericke, German** physicist, invented in 1663 the first electric generator. It produced static electricity by applying friction against a rotated ball of sulfur.

Stephen Gray, a British chemist, discovered in 1729 that electricity could flow. Scientists believed during the 18th and the beginning of the 19th century that electricity was a kind of fluid. In 1973, **Charles Francois de Cisternay DuFay**, a French chemist, divided this fluid in 'vitreus' (glass from Latin) or positive and 'resinous' or negative electricity. **Pieter van Musschenbroek**, a physicist and mathematician, first managed to store an amount of electric charge in a device called the Leyden jar. Independently, the same was managed by **E. Georg von Kleist**, German administrator and cleric. In 1752, **Benjamin Franklin**, an American printer and publisher, author, inventor and scientist, proved that lightning is an example of electric conduction. **Charles-Augustin de Coulomb**, French physicist, established a mathematical equation reflecting law in electricity. **Luigi Galvani**, an Italian scientist, experimented with electricity on animals. With further work in that field by **Alessandro Volta**, a physicist from Italy, it was constructed first battery as a source of continuous current. It was called voltaic pile. That was invented around 1800. There were many more very great minds who made significant contributions to progress of mankind in that field. I would lose my direction if I were to go into depth in this history. Perhaps to mention only one more name: it is **Henry Cavendish**, the greatest experimental and theoretical

English chemist, who did calculation of density of the earth with precision within 1 % of the currently accepted figure. He experimented with electricity, establishing that its intensity is inversely proportional to distance. A famous laboratory for physics was named in his honour as the **Cavendish Laboratory** where **Joseph John Thomson** discovered an electron. There was also a neutron discovered by **James Chadwick**. **Robert Rutherford** also started his work at the Cavendish Laboratory. In 1871, **James Clerk Maxwell** was elected as the first professor of the Cavendish. The laboratory was not completed until 1874.

Magnetism is a phenomenon which is the result of the motion of electric charges either through a conductor such as an electric wire or through space such as the movement of electrons in orbit around an atomic nucleus. In such situations where electric charges are on the move, they produce a magnetic field, which is perpendicular (at the angle of 90 degrees) to the direction of the charge movement.

If we have an electrical current going through wire then a magnetic field is created perpendicular to the direction of the charge in the wire going in a circle around the wire (Picture 2.04). If we put the thumb of the right hand in the direction of the current in the wire and wrap the fingers around the wire, they will point out the direction in which the magnetic field will go (Picture 2.05). The magnetic field is always marked as **B** field while an electric field is always marked as **E** field. An electric

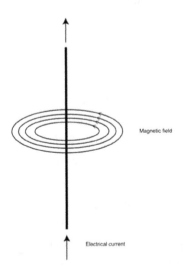

Picture 2.04

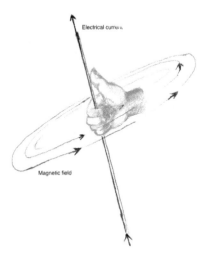

Picture 2.05

field goes in a direction from positive to negative charges. A magnetic field goes from North to the South Pole of its field. In the case of a simple single straight electric wire with an electric current and magnetic field circling around the wire, it is difficult to define and know where on this circle is North and where is the South Pole. If we, however, make a circle of the electric wire and switch electric current to go through such a circle, a magnetic field will be created where it is easier to locate the North and South Pole. Making a coil of wire long and shaped as a pipe (this is called solenoid) makes en electromagnet. Here such a pipe of coiled wire is wrapped around a long piece of iron. When electric current is switched through a solenoid then an electromagnet is created (Picture 2.06).

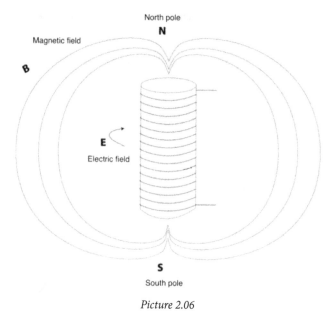

Picture 2.06

If we wrap the fingers of our left hand around the solenoid in the direction of the current of the circled wire, and make a thumbs-up sign, then the stretched thumb will show the location of the North Pole (Picture 2.07).

This is important as if the electrons in the circled wire change their direction then the location of the poles will change their sides as well. For example, if the electrons in the circled wire go clockwise, the North Pole will be up and South down. If electrons go anticlockwise then South will be up and North down (Picture 2.08). This is important in order to have a rough idea about magnetism.

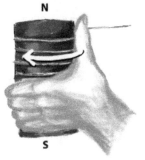

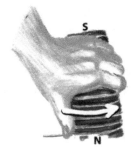

Picture 2.07   Picture 2.08

We know that in our world permanent magnets exists. This refers to a magnet which is already made as such by nature or to a material, which can be magnetised to become a permanent magnet. This kind of matter has a **ferromagnetic** property. Such a magnet has its magnetic field, which can affect another matter by attracting them.

These materials, which can be attracted by magnets, are called **paramagnetic.** These materials can align their atoms and attach to the magnet but cannot maintain magnetic property and when they are away from the magnet and its magnetic field they do not act as a magnet and cannot attract to each other.

Why do some elements have magnetic properties while others do not?

The answer is in the structure of an atom. Every atom has electrons orbiting around it. Every such electron creates a magnetic field by circling in its orbit around an atom. However, each orbit around an atom can have only 2 electrons; otherwise, the Pauli exclusion principle will be broken. The main point is that these 2 electrons circling around an atom do that in the opposite direction to each other: one circle clockwise and another in an anticlockwise direction. By doing so, one creates the North Pole at the location where the other will create the South Pole. This means that the magnetic field created by these 2 electrons will cancel each other out (Picture 2.09).

N   S   NS

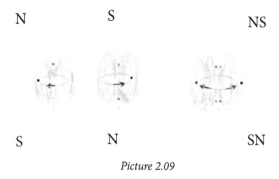

S   N   SN

Picture 2.09

Therefore, all those elements whose atom is made of a pair of electrons orbiting in each of his orbitals will not express its own magnetic field. If, however, there is only one electron in orbital or one electron in orbitals then the magnetic field of an atom will not be completely cancelled out. What location of North or South Pole of this particular atom will be, will depend on the rotation of the electron whose magnetic field is not cancelled out. In material with paramagnetic property, all atoms in this material have their random position. As each atom has a different position, the direction of their north magnetic fields is in all directions. As a result, we do not have a magnetic property of this material as a whole. However, when we bring this material in a magnetic field of a magnet, then all its atoms align in a direction south-north, reaching magnetic property, and start behaving as a magnet. Once a magnet is removed, their atoms again go to a random position with a subsequent loss of magnetic properties.

With iron, which has a ferromagnetic property, the structure of such material is divided into so-called domains. Each domain within the structure of iron material is aligned, expressing a strong magnetic field. However, each domain is differently aligned to each other so as a whole they cancel each other. Once this structure is brought in a magnetic field of a strong magnet, then each domain is aligned with a magnetic field. They tend to remain aligned so, after the magnet is removed, they remain a magnetic property or become a permanent magnet.

Just to mention that it is not only the circular movement of electrons in orbitals that create a magnetic field in atom but also the rotation of the electron around its own axes, so-called intrinsic spin. Such spin also has protons and neutrons in the nucleus and also creates a magnetic field. This is only for information and I will avoid going into this in detail.

An **electromagnet** is an example of a nonpermanent magnet, which is switched on when an electric current is switched on and switched off with switching off an electric current.

The history of magnetism dates from the time when a type of rock containing magnetite mineral was discovered. The structure of this mineral is a compound of iron and oxygen written as a chemical formula Fe_3O_4. It was found in Magnesia, the southeastern area of Thessaly in Central Greece. That is how it gets its name.

It was Thales of Miletus who first experimented with magnetism in around the 6th century BC. Around the same time, Indian surgeon Sushruta Samhita used magnetism in surgery.

In China around the 12th century, the Chinese used a lodestone compass for navigation. (A metal which has a magnetic property is called lodestone.)

The first significant progress in understanding magnetism was made in around 1600 thanks to **William Gilbert**, a physician and a personal doctor to Queen Elizabeth I. He was the first scientist to discover and state that Earth itself is a big magnet. With continued progress in this field and a number of discoveries and significant work done by many very important people in this field, the pinnacle of achievement was made by work done by James Clerk Maxwell. He basically unified magnetism and electricity, demonstrating that an oscillating current produces an electric field parallel to the direction of the current but also the change of current produces a magnetic field which is perpendicular to the direction of the current and electric field. Both electric fields, manifested as wave, and magnetic field as magnetic wave (both perpendicular to each other) propagate through the space perpendicular to the direction of propagation (Picture 2.10).

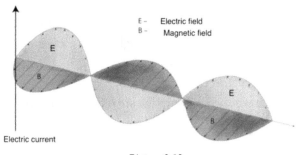

Picture 2.10

Maxwell further established that they travel at the speed of light and that electromagnetic waves are, in fact, what light is made of. It depends on their wavelength which part of the spectrum of electromagnetic waves they will be represented with. For example, if the wavelength of electromagnetic waves is in range from 10, 100 or 1000 metres then they will be represented as radio waves or, better to say, expressed and perceived as radio waves.

If their wavelength is at the centimetre scale they form or are called microwaves, which are usually in a wavelength of around 12 centimetres. If it is less than a millimetre or just below a micrometre, the electromagnetic radiation is called infrared, which is emitted by objects close to room temperature.

Light is electromagnetic radiation with wavelength between 0.4 to 0.7 micrometres. (Micrometre is a million times smaller than a metre of 1 x 10 on power of –6). Smaller than that, and up to 10 nanometres, we have ultraviolet radiation. Below 10 nanometres to 1/100 of a nanometre, they are called x-rays. With wavelengths below 1/100 of a nanometre, they are called gamma rays as the most powerful electromagnetic radiation with very small wavelengths and high frequency making them electromagnetic waves with very high energy and therefore the most powerful. Picture 2.11 shows radiation from the largest to the smallest wavelength.

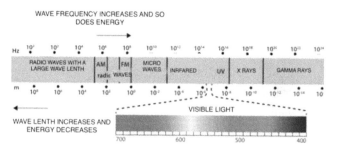

Picture 2.11

STRONG FUNDAMENTAL INTERACTION

Strong fundamental interaction presents the most powerful force in nature. Its effects are limited within the range of the atomic nucleus or to be more precise within the size of protons and neutrons. These are interactions between quarks, which keep quarks together within protons (2 up and 1 down quark) and neutrons (2 down and 1up quark). They are so strong within the radius of proton or neutron that they cross boundaries of protons and neutrons. This residual strong force or spillover is called nuclear force, which keeps protons and neutrons together in an atomic nucleus. Their range of dominance is within the distance, which roughly matches the length of 2 and a half diameter of portion. This is as far as spillover or residual force of strong nuclear interaction is concerned. It is important to make this clear distinction between strong nuclear interaction which acts between quarks in baryons (portions and neutrons) and its leftover or spillover called residual strong force or nuclear force.

The force carrier of strong nuclear interaction is gluon. It is exchange particle, which is located within hadron or baryon and acts as an exchange particle only between quarks. In order to get a rough idea or a slight understanding of this fundamental interaction, we should compare it with electromagnetic fundamental force or gravitational force. In electromagnetic fundamental interaction as well as gravitational one, the strength of these interactions or forces depends on the quantity of charges, positive or negative and mass quantity respectively. In case of electromagnetic forces or interaction, it is a photon, which serves as a carrier of this force, which is neutral.

In the case of strong nuclear interaction, its strength originates in a property called colour. This has nothing to do with original colour but it has some analogy to colour in the sense of a colour mixture and aim to get net of zero colour. What does that mean?

Well, in case of electromagnetic force and imbalance of charge will create electromagnet force moving an electron towards positive charges with a final aim to achieve zero charges. Such behaviours between positive and negative particulars are mediated by photons, particles that exchange this information telling positive and negative particles to rush towards each other. The final result will be zero charges. However, in electromagnetic forces we have only two kinds of charges or particles, positive and negative.

In strong fundamental interactions, which act on quacks in baryons, we have present 3 quarks or 3 particles. Each quark has its own colour, which could be either red, blue or green. These colours are interchangeable among 3 quarks within a baryon. It is not important how many times they change colours among themselves within a proton or neutron. What is important is that at any given time we have 1 quark, which is red, 1 which is blue, and 1 which is green within a baryon. It is important to always have this combination of colour within a baryon as the combination of red + blue +green gives zero net colour. This is in analogy to normal colours. We know that normal clours are part of the light spectrum which itself does not have colour or it could be referred to as a white colour. In essence, light itself can be defined as a zero net colour. We can do an experiment by making a paper circle and painting this circle 1/3 red, 1/3 blue and 1/3 green. We can then attach the centre of that circle to a pan and then start rotating the circle quickly around its own centre. In this situation, red, blue and green colours are not seen any more as separate colours. Instead, they will fuse into one colour, which will be white. We will have then zero net colours.

The gluon's responsibility is to make sure that this combination of colour always exists within a baryon.

Unlike a photon, which is neutral, a gluon is colour charged and in its interaction with a quark, it radiates colour itself.

The particular characteristic of strong fundamental interaction is that it gets stronger with the increasing distance. That is not the case with electromagnetic and gravitational interaction, as their strength gets weaker with the distance (their strength is inversely proportional to the square of distance between particles). The force of strong nuclear interaction can be compared to an elastic band attached to quarks which is relaxed when quarks are close to each other but if they move apart, the elastic gets stretched and pulls them back. In other words, gluons present within baryons are not felt or detected when quacks are close together. In such a situation the quark can behave like an independent particle. This effect is called asymptotic freedom.

David J. Gross, **H. David Politzer** and **Frank Wilczek**, American physicists, were awarded the Nobel Prize in 2004 for work in related strong nuclear interaction. They discovered that strong forces become weaker at smaller distances and that they become stronger as the quarks move apart.

Quarks come in triplets within protons and neutrons when a red, blue and green colour has to be present to give zero net colour. They can also come in a combination of quark-antiquark where one quark has a colour and antiquark anticolour, which brings again zero net colour. In reality, it is impossible to separate quarks on their own as they always come as three quarks within a baryon. If they pull apart from each other and reach the distance which is within the size of a hadron or baryon, the strong force reaches and remains at strength of 10 000 newtons. (Newton is the unit of force. 1 Newton is a force needed to apply to move 1 kg in a distance of 1 metre with an acceleration of 1 metre in a second per second.) That is the strength of the force, which keeps quarks together in a baryon. If we apply such an amount of work against this force to neutralise it in order to get free quark , we cannot get it as with that amount of work we can easily create particle-antiparticle pair. So that energy used with an intention to separate quarks will create new quarks, which will pair up again with the original one. This inability to get a free quark out of a baryon is called **colour confinement**.

3
ENERGY

We all have a rough idea in our minds what energy means and although we might not be able to accurately define energy, we more or less understand this notion of what we refer to when we talk about energy. To define it more precisely from the aspect of physics, we need help from physics. Physics defines energy as a capacity to do work. Work is a force used to move an object a particular distance or to displace an object at a particular distance. The equation for work in physics is therefore:

work = force x distance

Energy is therefore tightly related to work, as without energy, work cannot be done. In our everyday lives we know that the more physical exercise we do, the more energy we spend. If we additionally eat less or reasonably, then we will be able to keep ourselves fit, slim and healthy. This, of course, refers to middle-aged people who around that time usually start having a problem to balance intake of food or energy and the output of calories brought in. One of the reasons that middle-aged people are usually heavier or have more fat than when they were younger is a decrease of metabolism, which comes with age. Metabolism is regulated by the thyroid gland. This, however, is not a topic of this book and we can move away from this discussion as quickly as possible.

Energy is measured in joules.

1 joule is equal to 1 Newton (1 kg m/s on power of 2) times metre or:

$$J = \frac{kg\ m^2}{S^2}$$

S= second
M= meter

There is an old system of measuring energy, which is still in use when we measure the level of energy taken by food, and that is the calorie.

The calorie is used for two units of energy:

Gram calorie symbol of **cal**(small calorie) which is the amount of energy needed to raise the temperature of 1 gram of water by 1 degree celsius.

Kilogram calorie with the symbol of **cal** or **kcal**, which is the amount of energy needed to raise the temperature of 1 kilogram of water by 1 degree celsius.

A gram calorie is equal to 4.2 joules and this old system of energy measure is still used in chemistry.

A kilocalorie is equal to 4.2 kilojoules and is still used

to measure energy levels food contains itself. Otherwise, an official agreement is to use the International System of Units, which in the case of measuring energy is in joules.

There are many different forms of energy. We are familiar with many of them such as kinetic, which is the energy of a moving object; potential energy, which is stored energy (gravitational or elastic, for instance) and radiant energy which is carried by electromagnetic waves, electric energy and so on.

Energy can be transferred from one form to another. This transfer can be done by work or heat. I will elaborate more on these two ways of energy transferral, looking at these from a molecular level.

In Chapter 1, under the subsection of subatomic particles, we have outlined some conservation rules in relation to particles such as conservation of baryon and lepton numbers as well as conservation to charges of particles where we have zero net charges in a whole.

This rule is also applied to energy; but before stating this rule, which is the formulation of the first law of thermodynamics, I would like to introduce some concepts which might be helpful to make better sense of this rule.

For this purpose the whole universe can be divided into two compartments. On one side we have a numerous number of systems where each system makes one compartment. On the other side we have surrounding of the system, which makes another compartment. We are each individually one system which is separated from its surroundings by our skin. A football is another example of the system; one single biological cell is another system.

A pot of water prepared to be heated up for making soup is another example of the system and so on.

Every system has its own internal energy, which is a total amount of all energy, which this system contains within itself.

The total amount of all energy in the system or internal energy can be increased or decreased if there are transfers of energy between the system and its surroundings. If energy goes in the system from the surroundings, then the total energy of that system will increase. If it goes out, then the total energy of the system will decrease. This change of internal energy in the system is possible only thanks to the existence of its surroundings. These will allow transferral of energy between the two compartments.

The universe as a whole is a closed and isolated system. It does not have surroundings. It does not have outside or other compartments from where energy can be injected into the system and so increase the internal energy of the universe or where transferring energy from the universe to its surroundings can decrease the internal energy of the universe. As the universe is isolated with no outside, we have the first law of thermodynamics, which can be formulated as follows:

The internal energy of an isolated system is constant.

From there we can say that energy cannot be created or destroyed but only transferred from one form to another. As mentioned before, this transferral takes place by work or heat.

Looking from the molecular level, work transfers energy by using organised motion of the molecules.

Work is force multiplied by distance. When we kick a football, we inject or transfer energy into the football, which is manifested as kinetic energy of motion. The football as a system moves upwards against the force of gravity. In this process all atoms or molecules of the football move in the same direction, meaning that work use organised motion of particles, molecules.

Heat uses disordered motion of molecules to transfer energy. When we heat up a pot of water to make soup, we are doing this by applying thermal motion of the surroundings on the system. Disorderly motion of the molecules is called thermal motion. In the case of our pot of water, our hob starts producing disordered motion of the molecules, which stimulates disordered motion of the molecules of water inside the pot and in doing so, increases the temperature in the system where, with the help of heat, the energy of the system increases.

As stated before, energy cannot be lost or created but only transferred from one form to another.

Before it was kicked, the football had its own internal energy. Once energy was transferred to it by work, the football's internal energy increased, which was manifested as a kinetic energy. The football will go up but its movement will slow down by Earth's gravity. In one moment, for a split second, the ball will stop in the air. The whole injected energy will be now transferred to a potential energy. Potential energy is nothing more than stored energy (if we just remind ourselves of an electron jumping from an inner shell to an outer shell, absorbing a photon and gaining potential energy which is stored in the shape of a photon being absorbed). The ball will now head towards Earth, transforming potential energy to kinetic energy. (The same happened with an excited atom where an electron now goes back to the inner shell and emits that photon which was absorbed to go up to the higher shell.)

Based on the conservation of energy or the first law of thermodynamics, we cannot have or construct any perpetual motion. Perpetual motion means continued motion indefinitely with no external source of energy. A ball will not go upwards in the air unless it is kicked (external energy injected as kinetic energy). It will not continue to move upwards indefinitely as it will be slowed down by gravitational force and the friction it has in contact with molecules of air. In this friction, kinetic energy is reduced by being transferred to molecules of air. As this increases the movement of these molecules of air, their energy (referred to as molecules of air) is increased. In other words, a part of the kinetic energy of the ball is transferred in its surroundings or air by heat. So in order for the ball to continue to move upwards, the ball needs the constant input of external energy and, therefore, perpetual motion is not possible.

The internal energy of the system is the sum of all energy which is within the system. It is usually the sum of kinetic and potential energy. Kinetic energy is manifested with the presence of the movement of atoms, molecules. There are all sorts of movements, which take place at molecular level. Atoms can vibrate jiggles within molecular structures. Molecules can move randomly within the system. Electrons can jump on a higher level, having a potential energy, for example. In essence, every system has its own internal energy and as such, it can transfer this energy to its surroundings. Such transferral of internal energy to the outside of the system is called **thermal energy transfer.**

This can be achieved in three ways:

1. *By conduction*: when we hold one side of a wire while the other is heating up, we will quickly start to feel the heat as molecules or atoms of the wire transfer energy by touch to each other until it reaches the end of the wire we hold
2. *By convection*: where molecules move around transferring energy such as in the case of air where hot air moves upwards while cold goes down, creating circulation
3. *By radiation*: where internal energy is transferred by radiation or electromagnetic waves

Basically, all matter radiates energy as long as it has energy inside itself. It will stop doing so if it reaches a temperature of absolute zero, which is -274 degree celsius or 0 Kelvin. At this temperature there will be no movement and energy will be equal to zero. We should remember that an atom has its own energy due to its own structure as its electrons go around the shells. Each shell is at a particular energy level but when electrons go down to the inner shell, it loses or radiates this energy.

We also have protons and neutrons within a nucleus, which have some form of energy level. That is how gamma radiation is explained,which is emitted when protons and neutrons go back to ground state or lower energy level. However, as the temperature approaches absolute zero, electrons lose their energy levels, so do protons and neutrons. They radiate this energy away and coalesce with each other at the lowest energy level or ground state creating bosons; Bose-Einstein condensate.

We can ask the question, if energy is equal zero, where has all this energy disappeared to? Is that the break of the first law of thermodynamics? The answer is no, as we are decreasing the temperature in the system which has its surroundings. We are doing this by decreasing the temperature of its surroundings which makes the system radiate its energy out to its surroundings. In this process, the energy is taking away from the system and if the system is brought to near the absolute zero (we cannot achieve absolute zero) the energy of the system still exists but is transferred from the system to outside as heat and in doing so, the temperature of the system decreases while the temperature of its surroundings increases. Obviously, when the temperature of the surroundings increases, due to energy being radiated from the system, equilibrium will be established at one point. This means that the temperatures will equalise between the system and their surroundings. Once this happens, we need to find a way to cool down the surroundings again to allow further radiation of energy from the system until a temperature of zero is reached.

This also means that the whole process of cooling the system down to the temperature of absolute zero will take place in stages as the surroundings need to be cooled down again. To achieve this, we need a machine which will operate in that way. In other words, we need an extra energy.

We have just described the second law of thermodynamics, which can be formulated as:

Heat always flows from a hot object to cold and never from cold to hot.

This is one way of formulating the second law, but it is not complete. To complete it, we need to look at the first law which says that energy is always constant but only changes or converts from one form to another. While this is true, the initial form of energy or work used is never going to be completely transformed into another form of useful energy. It will always be a waste of energy as heat. If a car engine produces work or energy to move a car, this energy is never going to be completely converted to kinetic energy or motion of the car, but part of it will be lost as heat. In order to produce car motion or work, we always need to input more work into it as during conversion of our work to a different sort of work, we have a waste of energy as heat.

Work in (external energy) = work out (transferred energy) + heat (waste energy)

The general rule or tendency in the universe is that transfer from one type of energy always goes from the most useful kind of energy to the less useful kind of energy. This is where we come to the notion called entropy.

Before going to entropy, just to state that cooling the system to the absolute zero of -273 Celsius or Kelvin can never be achieved. It can come close but never completely, which is the **third law of thermodynamics.**

Entropy is the measure of the amount of disorder in a system. As the amount of disorder increases in the system, so does the amount of energy. However, *the amount of usefulness of energy is reducing.*

In the universe as a whole, entropy is spontaneously increasing and with it the usefulness of energy is decreasing. Energy is not disappearing but is becoming less useful every time the entropy increases.

Perhaps the best way to describe an increase of disorder or entropy with reduction of order and useful energy or work is if we are led by the definition of work at a molecular level where work is a way of transferral of energy by using organised motion of the molecules which is also a type of useful energy.

If we look at the phase state of the matter, taking the water as an example, we can state that ice (solid state of the water) has the lower entropy as water molecules are in a higher order. We can use ice to produce useful work, which is force multiplied by distance. We can throw sphere-shaped ice instead of a ball in bowling to knock over pins. When the temperature increases, the water melts, becoming a liquid, which has higher entropy end energy, but that energy is now less useful. We can still knock over pins with water, but we need more water to flow towards them. We can succeed in this only if we have a sort of tube which contains water and creates a condition for a stream of water to take place. If the flow of water is strong enough, it might knock over the pins. The energy of water will be less useful. Finally, with increasing temperatures, water will reach the higher entropy by evaporating in air. Energy will be higher but completely useless for the purpose of knocking down pins.

In essence, an increase in entropy inevitably reduces the amount of useful energy, which ultimately means that such a process will bring the universe to the end where the universe will consist of all the energy, as it does today, but all this energy will be useless.

We mentioned before that all matter has energy and can radiate this energy unless it reaches absolute zero. Once we have an absolute zero, then we have zero energy and zero entropy. But in any other situation, the system has internal energy, which can be transferred by radiation. It is done by electromagnetic waves. Waves are therefore the way of energy transferral. They are energy carriers. Electromagnetic waves are photons, which are carriers of energy or force in electromagnetic fundamental interaction between particles. We can simplify this and say that energy structure is an electromagnetic wave or photons while matter structure is quarks and leptons. Now it is the right place to give just simple general characteristics of waves.

WAVES – PHYSICAL CHARACTERISTICS

I will mention only a few physical characteristics of waves, which cannot by bypassed, as they are an important part of things, which make the universe.

A wave can be generally defined as an oscillation that propagates through a solid, liquid, air or space. Some waves can propagate only in a medium. An example is sound waves. They propagate through air, water and solid. All of these are mediums through which sound waves propagate. In this situation, the medium itself does not move but only the disturbance, which waves create, moves through the medium as a ripple. The best example is a wave created on the surface of water. If we have a leaf on the surface of this water, the leaf will move up and down the surface of the water as the wave passes beneath it but will not go away from this location as the medium, water in this case, does not move. As opposed to sound or water waves, which require a medium, electromagnetic waves do not require a medium and can travel in a vacuum. These waves are not created as a result of disturbances going through the medium, but as an oscillation of an electromagnetic field. As such, they go through an empty space unlike sound waves. So in space we will not be able to hear each other, as it is almost a vacuum with no medium. Light as electromagnetic waves, however, can travel through it.

Whether waves are sound, water or electromagnetic, they all have the same physical characteristics as follows:

Waves can travel as a singular wave, called a monochromatic wave, or as a number of waves, which are superimposed on each other. Examples are a light, which contains many different waves with different wavelengths, and frequency, which are superimposed on each other. This is called a spectrum.

Waves are composed of crests and troughs. Amplitude is the depth of a wave, or its precise size in length, from the base to the tip of the crest or trough of the wave.

Wavelength is the distance between the top of one crest to the next. Frequency is the number of waves in a unit of time.

The speed of a wave is calculated as a product of wavelength and its frequency. The speed of light is a constant and unchangeable. If we look at the equation of the speed of a wave which is:

Wave speed = wavelength x frequency

…then it is understandable that if frequency increases, the wavelength will decrease and vice versa. If we use numbers, it is easily demonstrated as follows:

$20 = 20 \times 1$
$20 = 10 \times 2$
$20 = 5 \times 4$

When waves reach the surface, they can reflect. The angle of refraction is equal to the angle of incidence. They can enter the new medium at a different angle from the angle with which they strike the surface. This is called refraction. This is due to the change of the speed of the wave in a different medium.

A spectrum of light when it hits a prism separates into its colours, of which it is composed. This happens as waves with longer wavelengths (red) refract or bend more than shorter wavelength waves (from red going towards a violet colour).

One very important effect, which can be observed from the standing point when waves approach us or go away from us, is the so-called **Doppler effect**.

What that means is what I hope I will be able to explain in the next few paragraphs.

When waves travel through the space it has constant wavelength and frequency. We can observe this if we travel along the wave with the same speed, the wave goes next to us or if the wave goes from right to left or left to right in our vision field. We will also observe the constant, unchangeable wavelength of a particular wave if it comes to us from a source which does not move.

We can imagine a boat, which has its engine switched on, producing a wave on the surface of the sea. If this boat does not move, we will observe wave after wave reaching us with the same distance between waves. In other words, the same wavelength.

The situation will, however, very much change if the boat is moving towards us. In that situation, the wavelength we observe will be shorter. The reason for this is that the first wave that left the boat will be followed by the next wave which will leave the boat at a position closer to us than was the position of the first wave leaving the boat. It is because the boat is moving towards us. As a result, the wavelength of these waves reaching us will be shorter and the faster the boat moves towards us, the shorter the wavelength will become.

The opposite will happen if the boat is moving away from us. Similarly, the first wave will be followed by the next wave which will be created by the boat further away than the first wave was created as the boat has meanwhile moved away. The faster the boat moves away, the longer the wavelength will be.

The Doppler effect applies to all waves involving light. The light spectrum consists of superimposed waves with different wavelengths from a blue colour being the shorter wavelength towards a red colour having the longer wavelength. When light reaches us from stars, which are moving towards us, we see light moving more to a blue colour, as the waves we see are shorter due to the Doppler effect. If light reaches us from stars which move away, then light will be perceived as a red colour or towards a red colour. This is called blue or redshift respectively.

At the beginning of the 20th century, it was noticed that galaxies are seen to be in redshift spectrum, indicating their movements away from us.

In 1929, **Edwin Hubble,** an American astronomer, published a paper which demonstrated the relationship between redshift and the distance of galaxies observed, showing this relationship to be linear. It means that if a galaxy is twice as far away as another, its redshift is twice as large. Therefore, it was established that redshift is directly proportional to its distance.

The Doppler effect can tell us if stars are moving away or approaching. If we perceive light from a star as blueshifted on one side and redshifted on the other, that means that this star or celestial body we observe is rotating.

There is a very important part of study of light within physics called **Spectroscopy**.

Spectroscopy studies the relationship between matter and radiation energy. It is particularly important for astronomy as it helps to analyse the structure and composition of stars, which are far away and out of reach. It can help to determine the temperature of the star as well.

It was **Gustav Kirchhoff**, a German physicist, who made a significant contribution to the understanding of spectroscopy. He identified three types of light spectrum:

1. A continuous spectrum, which is produced by a *hot solid object*. He called it black body radiation in 1860.
2. Emission spectrum characterised by spectral lines produced by *hot gas*. These lines will be in light spectrum in the particular colour of the spectra depending on the energy levels of the atom in gas.

Basically, when gas is heated up then it gets energy

impute or photons for its electrons to jump at higher energy levels or shells. Electrons then go back to ground state and emit light, which comes as a line within the spectrum. The lines are at a particular part of the spectrum, which corresponds to a specific energy level of this gas, or atom of this gas. Therefore, the composition of that gas or what elements are present there can easily be identified. (Oxygen will give a different line in the spectrum than nitrogen, for example.)

3. Absorption spectrum are black lines in the spectrum, which we can see when a continuous spectrum from a hot solid black body goes through *cold gas* which is between a black body which radiates continuous spectrum and us who observe. In this situation, the cold gas absorbs the part of the spectra that corresponds to the energy levels of the atom of the gas. In doing so, it leaves black lines on the spectrum as this part of the spectrum is now missing, being absorbed by gas.

Absorption spectrum can help us to identify what sort of elements are present on the surface of the Sun, for example. Namely, the core of the Sun, which is very hot, emits a continual spectrum of radiation which goes through gas on the surface of the Sun on its way to us. Also, the surface of the Sun is hot; it is cooler compared to its core. The gas or elements making its gas absorb part of the spectrum of this radiation in a particular place which corresponds to a particular energetic level of the atom of a particular element. These will appear as black lines in the spectrum as part of this spectrum is missing. It is absorbed by atoms where election jumps to a higher energy level. This is specific for every element which is helpful to identify the composition of stars despite their distance, which is out of our reach.

At this point, I should be elaborating on a particular kind of radiation, which comes from black body. It is important as thanks to this type of radiation Albert Einstein and Max Planck have come to the idea that energy carried by electromagnetic waves is delivered in packets of energy called photons. Precisely, Einstein came to this idea when he studied the photoelectric effect, while Max Planck came to this idea when studying black body radiation. This opens a new chapter in physics where quantum mechanics was developed which enhances progress in science and technology.

Black body radiation was also crucial in our understanding of the origin of the universe. It was thanks to the physical behaviour of black body radiation that predictions were made about microwave background radiation whose existence supports the Big Bang theory. It was **George Gamow** who predicted cosmic microwave background radiation in 1948, but it was not discovered before 1965 when this finally consolidated the Big Bang model as the correct theory of universe origin.

I will leave this for now, as it would be more appropriate to talk about black body radiation within a chapter where I try to explain or rather give my version of how I understand the Big Bang theory from the point of view of a layman such as I.

Finally, just to touch once more on conservation of energy and mass in light of mass energy equivalence as expressed by Einstein's equation:

$$E = mc^2$$

At the end of the first chapter, I outlined that mass can be changed into energy and vice versa using an example of binding energy. In such cases, we are talking about a slight variation of the mass or mass reduction. However **c** squared is a big number, which shows that even with a very small amount of conversion of matter to energy, we get release of a huge quantity of energy. That is why release of nuclear energy is so enormous and powerful.

In the case of fusion of hydrogen nucleus in nucleus of helium, the binding energy, which was released from the fusion of protons, is so powerful outward that it balances strong gravitational forces of the enormous mass of a star which moves inwards. It is the reason the star is not collapsing on itself. The release of energy in nuclear fusion will continue to take place until iron is created. Any element after iron cannot be created by nuclear fusion within the star as in this process we now need to bring energy in for a new heavier atom to be created.

In a process of fusion, where the new element is created, the mass of nucleus of the new element or atom is slightly smaller than the sum of masses of nuclei of atoms of elements which were fused. This is due to binding energy, which is released when fusion takes place and which comes from reduction of masses of atoms involved in fusion. This process is repeated until iron is formed. From this moment, a further fusion process cannot take place, as energy needed to fuse these iron atoms is higher than binding energy relisted to bind protons and neutrons in an atomic nucleus. The mass of the newly created atomic nucleus will now be higher than the sum of the masses of the iron nuclei that took part in nuclear fusion. In that sense, the energy would not be released but it will be absorbed in order for further fusion to take place. That is why fusion of iron in a star is the moment when the star comes to the end of its own existence.

To go back to the law of mass-energy conservation, it can be stated that mass-energy is constant in the universe where it is possible to achieve some slight changes in a mass where some energy can be released from the mass, reducing the amount of mass but increasing energy which can be again incorporated or absorbed by mass locking this energy away in mass by increasing this mass again, but altogether giving a constant amount of mass-energy within an isolated and closed system.

As already mentioned, only a very small amount of mass can be converted to energy unless we have matter combined with antimatter where all mass involved is converted into energy.

4
DARK MATTER

Dark matter makes up about 23% of the universe. Its existence is confirmed by an indirect method as dark matter does not absorb or reflect light and, as such, cannot be seen. It does have a mass and, as such, it is affected by gravitational forces as well as exerting this force. It is thanks to this that the existence of dark matter was identified.

Before going to this it would be good to explain how we see things around us.

First of all, we see nothing in a closed dark room with no light on. To see things, we need light or photons. But what is specific in light that allows us to see things around us?

To answer this, just recall that light is composed of various electromagnetic waves with a different wavelength, which are superimposed on each other. If we defined only one line of electromagnetic radiation, which has a particular wavelength, for instance, length, which is in a spectrum or red colour, as one light beam, then daylight is composed of various beams superimposed on each other in one light beam. Each such light beam has a different wavelength from the shortest one within a visible light spectrum which is violet colour and blue colour and going via blue, green, orange, yellow, to red colour with the longest wavelength within the visible spectrum of light.

We know again that electromagnetic waves are delivered in packages called photons. Travelling at the speed of light, photons after photons are experienced as a continuum in the way of a light beam. It is the same as a jet of water where so many droplets of water go at such a speed one after another that this is experienced in a continuum in the way of a water jet.

One chunk of beam which corresponds to the size of a photon has insight itself superimposed various number of photons, with each one having a specific wavelength representing a particular colour of the light within a visible spectrum of the light. We know that many photons, which are superimposed on each other, can do that as they are bosons and, as such, are not subject to the Pauli principle of exclusion. In other words, it can be as many as possible number of photons at that same place and the same time.

We are now coming closer to understanding how we see things around us.

When beams of light or plenty of superimposed photons heat the surface of an opaque material, then plenty of these photons are absorbed by opaque material. How many and what kind of photons will be absorbed depends on the atomic structure of a particular material. If the atomic structure of material has electrons arranged in shells around the atom in such a way that there is the possibility left for electrons to absorb photons with a particular wavelength (say red colour), to jump on the outer shell, then that photon will be absorbed. If the atomic structure of an atom allows photons of all colours to be absorbed except the green colour, then that photon will bounce off or reflect and come to our eyes. As all the other colours are absorbed, we will detect this material or object as green-coloured, having only the photon of this colour being reflected and coming to our eyes. If all light is absorbed, we will detect this object or see it as black- coloured. If the colour of all light is reflected back, we will see this object as white-coloured. Fully opaque objects are mirror and black body . A mirror completely reflects light while a black body completely absorbs light.

Glass, however, has an atomic structure where the arrangement of electrons does not allow any space for a possible absorption of any light photon as there is no outer shell on which to jump, where a photon could be absorbed. It does allow photons of light to go through it completely with no interference or no scattering, making glass transparent.

Photons can scatter or hit or collide with matter and free electrons when they come into contact with them. If photons, which come in matter, scatter with electrons within the matter and do not come out, then this material is perceived as opaque. That was the case with the universe when the temperature was so high that matter existed in a plasma state with protons and electrons being separated. The electrons were moving at relativistic speed, close to the speed of light. They were on their way to photons, causing them to collide with the electrons or scatter all the time. That is why the universe was opaque then. When the temperature reduced to about 3000 K around 38000 years after the Big Bang, then the speed of electrons reduced enough for electrons to be captured by the atomic nucleus in atoms. Photons did not scatter with electrons any more and were free to go away in all directions. The universe then became transparent.

Photons do not scatter or hit each other as in this case we would not be able to see.

Obviously, the above explains physical phenomena as to why we can see things as we see; it would not be enough if we did not have eyes. Trees cannot see as plants do not have eyes.

Dark matter does not absorb, does not reflect light and it does not have any interaction with ordinary matter. That is why it cannot be seen. It can, however, be detected indirectly due to the mass it possesses which is subject to gravitational force.

There are two main ways of detecting or, better to say, two ways how dark matter was detected. The first is **orbital velocity** and the second is **gravitational lensing**.

The **orbital velocity** of a celestial body depends on gravitational pull and distance from an object or another celestial body, which exerts gravitational attraction. Just to recall that gravitational force between two bodies is inversely proportional to the squared of the distance between these bodies. Therefore, the closer the object is to the object of gravitational pull, the faster it rotates around this object. Mercury rotates faster around the sun than Venus. Venus rotates faster than Earth. The further away a planet is from the Sun, the slower it rotates around the Sun. Pluto has the slower rotation around the Sun.

Now, let us look at it in a slightly different way and imagine Mercury rotating around the Sun at a constant speed higher than any other planet due to a strong gravitational pull from the Sun. It stays in orbital motion due to a strong gravitational pull from the Sun. But now, try to imagine that the mass of the Sun has suddenly decreased or almost disappeared. Then there would be nothing to keep Mercury in such orbital motion. Instead, it would continue to go away from the solar system in a straight line.

Another good example is if you have experience with fishing. In the place where I come from, people used to catch fish by using a fish hook which was attached to a very long piece of string. It was no proper fishing device and was composed only of a long piece of string with a fishing hook at the end. The fishing was usually done from the shore. It was important to throw away the fish hook as far as possible into the sea. The best way to do that was to rotate the piece of string in the air with the fish hook at the end of it. Once they achieved good rotational movement of the fish hook, they would let go of the piece of string from their hand and the fish hook would fly away to sea.

The fisherman rotates the fishing hooke

He throws it and the fishing hook flies into the sea 4

Picture 4.01

This is very similar to the gravitational pull of the Sun or any object keeping another object in orbital motion around it. Once this gravitational pull disappears, like letting go of a piece of string in the above figure, the rotating object will fly away, as the fish hook did. With this logic in mind, we can understand how dark matter was discovered.

In fact, this is how the presence of dark matter was first detected in the early part of the 20th century.

It was **Fritz Zwicky**, Swiss astronomer, who discovered dark matter in 1933 when he studied the Coma cluster of galaxies. (This is the cluster of galaxies located in the constellation of Coma Berenices. The Coma Berenices constellation is next to the Leo constellation which can be found in the sky when we look at night from March to May. This refers to the Northen hemisphere, which means only part of the sky visible from the north part of Earth, Europe, North America, for example. See Puiture 4.02.)

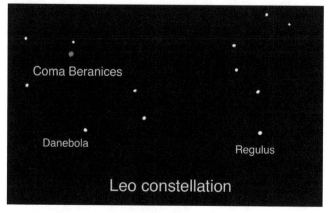

Picture 4.02. The Leo constellation can be easily seen with a head looking like a reverse question mark. Once you find the Leo constellation, it is easy to find the Coma Berenices constellation.
The Coma Berenices constellation has the Coma cluster (blue dots) next to it. Coma Berenices in Latin means Berenece's hair. It is mythology about Berenice II (born 269-died 221 BC) who married Ptolemy II Eurgetes, Macedonian King of Egypt, around 245 BC. When Ptolemy went to revenge the murder of his sister in Syria, Berenice promised to cut her long hair for his safe return. He returned safely but, as her hair was not found where it was stored following cutting, the court astronomer said that the goddess was so pleased that she decided to put her hair in the sky where it formed the Coma Berenices constellation.

Zwicky calculated the speed at which galaxies rotate around in clusters and the amount of matter available as seen by the luminosity of present stars. He found out that there is a huge discrepancy between the speed and the amount of available stars. It was only around 10% of visible mass, which if it was the only mass available, it would not hold galaxies to go around at such speed. In other words, if there was not some 90% invisible mass present, then there would not be enough gravitational pull and galaxies would fly away from each other.

In 1970, **Vera Rubin** and **W. Kent Ford** observed the same phenomena where mass of the stars in a galaxy only had 10% of mass needed to keep them orbiting at the speed they did around the centre of the galaxies. It is 90% of invisible mass which kept them going with such speed or otherwise, these stars would fly away.

Gravitational lensing is another important phenomena, which happens as the result of the gravitational pull large mass has on light. Einstein made calculations and accurately predicted the degree of banding of light, which depends on the amount of mass of the object the light passes near to on its way.

These phenomena can be observed when a massive object is between stars and us, which are far away. The light travelling from the star can be seen although it is behind a massive object and this is due to bending of the light due to gravitational pull as the Picture 4.03 shows.

Astronomers have observed that a number of galaxies in the distance are subject to gravitational lensing in some parts of the sky with no visible massive

Picture 4.03. Bending of the light due to a star (star is the circular object). Position B shows where the star light from a distant star would be seen if there was no massive star on the way on the light. Position A is where this star should not be seen when a massive star (circular object) is in front. However, due to light binding as it is pulled by gravitation of the star, that star is seen displaced at a place shown by the interrupted line.

object between the galaxies and observer. It was therefore the only possible conclusion for such phenomena that the gravitational lensing was caused by invisible mass, which will be dark matter.

With current technology, experts and people working in that field are able to identify, thanks to gravitational lensing, the exact place where dark matter is situated in the universe. The final result is amazing. It shows the arrangement of dark matter as a web throughout the universe. Whenever there is dark matter, there is a number of galaxies. Dark matter serves as scaffolding upon which ordinary matter is organised in stars and galaxies. It looks like the whole universe is built up in the way everything is built up around us, starting from single cell organisms to mammals, including us. If you research cell structure, you will find out that a cell has its own cytoskeleton (numbers of microtubules), which is the cell skeleton around which the cytoplasm and cell are organised and built. Animals, including us as the most developed species, have their skeleton, around which muscles and all the rest of the body is built. If we look at any building, we see that, in order to be built and maintain its structure, it needs to have a strong support of still wire or its own skeleton. The same logic can be applied to the universe. Galaxies are built around a skeleton of the universe, which is presented with dark matter.

My own illustration below is my attempt to show how this looks. As stated in the preface of this book, there are beautiful images and illustrations which are easily accessible online. I was tempted to use some of these as examples in my own book. However, as I do not know how to obtain permission to use them, I decided to make my own illustrations, but I would encourage you to have a look at the beautiful examples available online.

Dark matter like a spider's web with galaxies (blue dots) built on it. Pleas look at the Internet for these images, which are beautiful there.

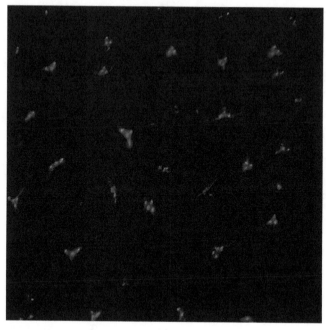

Picture 4.04

Over the last two decades, astrophysicists have focused their attention more on the existence of dark matter. They have tried to find out what the dark matter is made of. As far as I am aware, nobody has managed to identify the nature of dark matter. There are existing theories, which I will outline here with no intention of going into too much detail.

Ordinary matter (making up planets, stars, and galaxies, involving us) is made of baryons. Baryons are protons and neutrons. Electrons are around 2000 times smaller than protons and are somehow dismissed. That is why ordinary matter is usually referred to as baryonic matter. Ordinary matter makes up around 4 % of the universe.

The existing theories regarding the nature of dark matter are that it can be made of:

1. Baryonic matter
2. Non-baryonic matter, which again can be divided into

    Hot dark matter
    Cold dark matter

1. **Baryonic matter** can make up to 15 % of ordinary matter. It is suspected to be a cloud of helium within galaxies.

    MACHOs or massive compact halo objects are also suspected of being a baryonic part of dark matter. It was suspected to be a primordial black hole created during the Big Bang, faint red star or brown dwarf. I will give some details about the characteristics of each of these objects, or rather celestial bodies, in Part 2 of my book.

2. **Non-baryonic matter** is a different kind of matter which does not interact with other particles in the sense of the four fundamental interactions which exist among particles of ordinary matter, except for gravitational interaction.

    We know that the temperature of the system or a particular matter depends on kinetic energy or speed of movement of molecules or particles within the system or a particular matter. The quicker molecules, particles move, the higher the temperature of the system or a particular matter is.

With this logic in mind, we can understand further division of non-baryonic dark matter into hot dark matter and cold dark matter:

a. **Hot dark matter** should be composed of particles which move fast, achieving relativistic speed (close to the speed of light). It is suspected that this kind of matter is made of <u>neutrino</u>s.

b. **Cold dark matter** should be composed of massive particles, which move slowly.

WIMPs or weekly interacting massive particles are suspected to be the nature of cold dark matter. Within this, there is the theory of the existence of so-called super symmetric particles. It is popularly called SUSY. Namely, according to the Standard Model theory, particles have partners that differ in spin of ½. A photon will have its partner photino, for example, which could be a particle of dark matter.

These are current theories or speculations as to what might possibly be the nature of the dark matter, but so far there is no evidence to support or clearly identify its nature.

5
DARK ENERGY

Around 73 % of the universe is composed of dark energy. Roughly speaking, dark energy can be defined as an increase in space between galaxies, and that increase is accelerating. It is the repulsive force, which counteracts gravitational force.

It was Albert Einstein who first acknowledged the existence of dark energy but for the 'wrong reason'. Namely, at the time when he worked on the theory of relativity, the universe was considered to be static. It was supposed that the universe was not expanding, but not contracting either.

When we have gravitational force keeping the matter of the universe together in the shape of galaxies and clusters of galaxies, then it is difficult to imagine that the universe will remain static as gravitational force will pull matter together and the universe will collapse on itself. The universe was at that time considered to be static, so Einstein introduced the **cosmological constant**, in 1917.

It is repulsive force and the opposite of gravity with the exact same force as gravity but in the opposite direction or expansion rather than contraction due to gravity. As there is the exact amount of force but in the opposite direction then it is easy to understand that in such a case the universe will remain static. In every book or documentary regarding this topic, you can read or hear how Einstein commented on the introduction of the cosmological constant as his 'biggest blunder', once he found out that the universe is expanding. As we all know, it turns out instead to be one of his biggest achievements and predictions.

However, it was not until 1998 that dark energy was discovered thanks to work by two international teams, one involving **Adam Riess** and **Brian Schmidt** and the other involving **Saul Perlmutter**. They all shared the Nobel Prize in 2011 for discovery of dark energy.

Both teams used luminosity of type 1a supernovas, thanks to which they were able to establish distances between galaxies.

Type 1a supernovas have the same intensity of light or luminosity when they explode whenever they happen in the universe. Because of the exact amount of luminosity it is possible to determine the distance between galaxies when compared to the degree of brightness *type 1a supernovas* have at the particular galaxy whose distance is known and the brightness of supernovas in faraway galaxies.

How do they do that?

Very roughly speaking, it is known that brightness of light decreases with distance and that the decrease takes place gradually and proportionally to the increase in distance. The equation for it is as follows:

$$\text{Brightness} = \frac{L}{4 \pi R^2}$$

Brightness is flux or amount of light per unit of space while L is the whole light emitted in all directions, which travels away from the source as a sphere. This is actually formula which can be used to calculate the distance of light from us or distance of a particular type of star from us.

So far, a number of different stars have been identified which emit light with particular intensity. According to the degree of brightness a particular star produces, they are categorised or classified in several or precisely seven type of stars. A particular letter is given to a type of star with a particular brightness in order of decreasing temperature, and decreasing brightness as: **O**, **B**, **A**, **F**, **G**, **K**, and **M**.

O and B type of stars are very bright but uncommon while M stars are common but dim.

You can easily memorise the types of stars according to their brightness from higher to lower intensity by using the mnemonic 'Oh be a fine girl, kiss me'.

So if we find a G type star, as an example, in a faraway galaxy, but her brightness is less than that from a G type of star, which is closer to us with known distance, then we can calculate the distance of this star from us thanks to the degree of reduction of her brightness. When we know the distance of this star, then we know the distance of this galaxy, as this star is located in this particular galaxy.

Once we know the distance of particular galaxies then we will have the exact intensity of light emitted by *type 1a supernovas* occurring in this galaxy. Now when we know the distance of this galaxy and intensity of light of the supernova at that distance, we can compare this brightness with the same type of supernovas observed in a galaxy far away. Depending how much fainter this supernova from the closest one with a known distance is, we can calculate how far away is this galaxy.

Using this method, both teams of astronomers concluded independently that galaxies with supernovas at the distance when the universe was two-thirds of the current size are much further than they should be. It was because light from *type 1a supernovas* was fainter than it should be.

Type 1a supernovas are very helpful in measuring distances between galaxies. This was very helpful in the discovery of dark energy. Basically, by measuring distances between galaxies using this method, it was clear that universe expansion is accelerating which was due to the repulsive force of dark energy.

Type 1a supernova is used as a standard candle for measuring distances. Independent of this, another technique for measuring distances in space can be used, and that is baryonic acoustic oscillation.

Baryonic acoustic oscillation is used as a standard ruler

to measure distances. The distance, which it applies to, is a constant and is around 470 million light years or 490 million light years. I find different figures for it in different books.

I do not think it is very important to know precisely the value of this distance. What is important is to get a rough idea that this distance is a constant and is the largest yardstick used in astronomy so far, almost close to half a billion light years but not quite.

Type 1 a supernovas are those which are believed to have originated from a binary system consisting of a moderately massive star and a white dwarf. A white dwarf is a dead star with a large mass density. As such, it sucks into itself material or mass from its massive companion. The reason for this is that its companion is too close to white dwarf to escape its gravitational pool. However, getting the matter from a massive star, white dwarf is increasing in size. Once its mass grows above the Chandrasekhar limit of 1.44 solar mass, it will end in a massive thermonuclear explosion or type 1a supernova.

Subrahmanyan Chandrasekhar is an Indian-born American astrophysicist who won the Nobel Prize together with **William A. Fowler** for key discovery in relation to later evolutionary stages of massive stars.

He determined that any star remnant with a mass bigger than 1.44 times the mass of the Sun cannot exist as a white dwarf but instead blows off in a supernova explosion.

Baryonic acoustic oscillations is a notion referring to quite exciting progress made relatively recently in the field of astronomy or astrophysics.

Two Russian astrophysicists **Rashid Sunyaev** and **Yakov Zeladovuch** predicted their existence in 1970 but the oscillations were not seen until 2001 when balloon-based microwave detectors were constructed.

Baryonic acoustic oscillations refer to acoustic waves or ripples, which were created and travelling through baryonic matter at a very early stage of universe creation.

They were a result of opposite forces of gravitational attraction due to the enormous mass density of baryonic and dark matter, which was pulling matter inwards and the pressure of photons and electrons pushing it outwards. Dark matter, however, interacts with other particles only through gravitational fundamental interaction. It is subject only to gravitational force and therefore stays in the centre of the sound wave. Baryons and photons due to pressure created travel in the shape of a spherical sound wave outwards.

We should not forget that this is an incredibly hot and dense plasma state of the universe in the first few hundreds of thousands of years of its creation or after the Big Bang. After around 380 000 years since the Big Bang, when the universe cooled down to around 3000 K, the speed of electrons was reduced enough to be captured by a nucleus of hydrogen and helium to form the first atoms. The universe was not in a plasma state any more. Matter has organised in atom structures of mostly hydrogen and helium. The universe was not any more opaque but became transparent. That was a crucial moment as then photons suddenly got free and escaped, moving at the speed of light in all directions. That was uniform radiation of photons with frequency in spectrum of visible light. The remnants of this uniform radiation are today detected as cosmic microwave background radiation. The reason that this is not detected in spectrum of the light is the fact that the universe has stretched enormously since then, and with it waves within photons going from short wavelengths within the spectrum of visible light to long wavelengths within the spectrum of microwave radiation as we perceive it today.

In relation to baryonic acoustic oscillation, we have to go back to the moment when photons decoupled from matter around 380 000 years after the Big Bang. Up to that moment the pressure between baryon and photons was pushing both of them outwards. However, when photons decoupled from matter, this relieved the pressure. Photons then escaped while baryons remained fixed at this radius. This radius is often referred to as the sound horizon or the zone of the last photon scattering. The pressure created by photon and baryon interaction does not any more drive baryons outwards as this pressure has now gone with the escape of photons.

To understand this better we can imagine boiling water in a pot with a lid on top of it. When water starts to boil and the lid is still on, the pressure of molecule of water interaction will push the lid, making it dance and creating sound due to the increased pressure driving the molecules of water outwards. Once we remove the lid, the pressure will decrease and bubbles of boiling water will for a moment goes down or inwards.

Similarly, when photons escape, then the only remaining force for baryons is gravitational force pulling the matter of baryons towards the centre where there is dark matter which did not move from there all the time. This creates an area of more dense matter at the original centre and at the sound horizon.

We know that gravitation depends on mass density. As we now have zones where the number of baryons is more per unit of volume than in other places, then these will attract more mass than others and would eventually create stars and galaxies around these areas. At the same time, these places will have a higher temperature than others, as the number of baryons is higher with more interactions among these particles. They will therefore radiate more energy, which will cause radiation of energy not to be so uniform. This will be imprinted in cosmic microwave background radiation, which is detected as anisotropy of CMB or slight changes in temperature at some places on a map of CMB. Well, this increase in temperature corresponds to the area where baryonic mass density was higher than in other places. It corresponds to the ripple of baryonic acoustic wave oscillation, which was the seed for galaxies' formation. These distances between slightly higher temperatures are constant and not changeable. It is close to half a billion light years and corresponds to the distance of the first galaxies being formed around a billion years after the Big

Bang. The reason this distance is constant is that they are perceived from CMB, the signals which we receive now about something which took place around 10 billion years ago. It is therefore a snapshot of the time around 10 billion years ago.

When we look at our photographs from childhood we see how we looked at that time. The photograph is a snapshot of the time when we were children. The particular time of our age is arrested by the photograph made at that time. In other words, the picture in the photograph does not get older but remains a reflection of how we looked as children at that time.

Now we will go back to our universe and imagine one hypothetical scenario where the universe is a constant size, not expanding and not contracting. We looked at distances between galaxies, which are at different distances from us. For example, we looked at galaxies A and B, which are at a distance of 8 billion light years from us and A1 and B1, which are at a distance of 5 billion years from us. A1 and B1 are galaxies closer to us as it took 5 billion years for light from them to reach us. We can make a comparison of the distance between A1 and B1 galaxies with the distance between A and B galaxies, which are 8 billion light years away. We should not forget that the light of these galaxies we receive now is the snapshot of the time of how they looked at 5 or 8 billion years ago. It means that the universe has moved forward in time 3 billion years from the snapshot received from galaxies A and B to the snapshot received from the galaxies A1 and B1.

As we consider the case that the universe is not expanding or contracting, then baryonic acoustic oscillations, or the distance between them at the time when the first galaxies were formed, should correspond or exactly fit the distance between A and B galaxies as well as A1 and B1 galaxies.

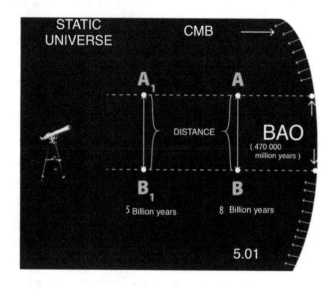

Picture 5.01

Now we will consider the universe, which is expanding, as it is the case in reality. In this case, the distance between galaxies A and B is larger them BAO (baryonic caustic oscillation) as the time of the universe moved forward 2 billion years. CBM is a snapshot of the time 10 billion years ago while A and B galaxies is a snapshot of the time 8 billion years ago. 10 – 8 =2. During this time of 2 billion years the universe has expanded, and with this expansion the distance between the two galaxies A and B has expanded so that BAO or distance, which was between galaxies first created 10 billion years ago, cannot fit exactly. The same applies to the distance between galaxies A1 and B1. The universe has expanded so much, and with it the distance between A1 and B1 galaxies, that we can now fit 3 or 3 and a half BAO within this distance.

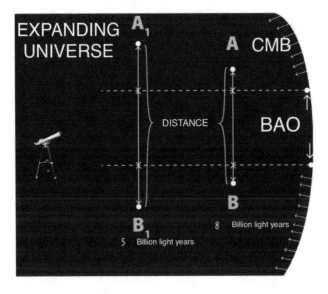

Picture 5.02

With current technological progress, the discovery of baryonic acoustic oscillations, for instance, and the general improvement of devices which help us to observe the universe more accurately, we are now able to make a very confident conclusion that the universe up to 9 billion years of age was slowing down in its expansion. It was because matter dominated the universe. The universe is about 14 billion years old. In many sources of information, it is stated that the age of the universe is about 13.7 billion years. Roughly, it can be said that it is 14 billion years.

From the time when the universe was 9 billion years old, the expansion of the universe began to accelerate. This acceleration of universe expansion in the last 5 billion years is due to dark energy taking dominance over matter.

How do we know that this expansion is due to dark energy?

It is actually a logical conclusion that there must be something or some form of energy in the empty space, which increases the space between galaxies. The names *dark matter* and *dark energy* are given as it is not yet clear what dark matter is made of or what is the source of dark energy.

To understand it better we can imagine throwing a ball with a powerful force (a cannon ball, for instance, from a cannon) upwards against the gravity of our planet. The initial speed of the ball will be high but as it travels away from the Earth's surface it will slow down. It will eventually reach the pick of the distance between itself and the Earth and will start travelling back to the Earth, being pulled by the gravitation of the Earth. We have therefore an expansion of the space between the ball and the Earth initially. This expansion would slow down due to gravity, which would eventually reverse the ball's direction with the ball now moving towards Earth with a subsequent reduction of space between the ball and the Earth.

Similarly, it was expected that the universe expansion should be decreasing until the maximum expansion is reached, after which the universe will reverse back and start contracting, ending up in a 'big crunch'.

For up to 9 billion years of its existence the universe was slowing down but then it started to accelerate significantly which is the opposite of what is expected when we take into account gravitational force as the most important fundamental interaction between particles at great distances.

Therefore, the recent discovery of the fact that the expansion of the universe is accelerating could be due to two possible mutually exclusive reasons, which are:

a. Either the concept we have about gravitation is wrong or we are missing something there, or
b. There must be some energy of empty space which is doing it. This means introducing a new ingredient to make sense of the current theory.

In history we had situations where we had to introduce new ingredients to make sense of an existing theory. An example is the Ptolemaic geocentric system. Claudius Ptolemy was a Greco- Roman astronomer and mathematician who lived in the first half of the 2nd century AD. At that time, the logical conclusion for the observation of the universe was that Earth is the centre of the universe and that all celestial bodies go around Earth. However, planet movements (those planets which were known at that time) were unusual and could not be explained by a geocentric model. In order to do so, there was a need to introduced a new ingredient or throw away the geocentric theory as not plausible. Ptolemy introduced a new ingredient instead. It was called epicycle, which explained unusual movements of the planets and fitted into an existing geocentric theory so well that the theory remained a logical and plausible explanation of astronomical phenomena until the Copernicus heliocentric system.

Similarly, we can reject the theory of gravitation or introduce a new ingredient, dark energy, to fit the current model and understanding of the universe.

At the present time of understanding of phenomena in the universe, the nature of gravitational force remains very logical, plausible and cannot be rejected.

We therefore have to introduce a new ingredient or dark energy to make a sense of the current observation.

Currently, we do not know what exactly the dark energy is. Like there are many speculations as to what dark matter is composed of, we have speculations as to what dark energy is made of. Two main candidates are:

1. **The cosmological constant**
2. **Quintessence**

The cosmological constant, which is also written as the Greek letter *lambda*, was used by Albert Einstein to refer to repulsive force, which is the opposite to gravitational force.

The exact nature of dark energy or the source of this energy is not yet understood. There is a mathematical calculation which provides a logical explanation from a mathematical point of view and, to some extent, I might be able to understand it. I am, however, not a mathematician or physicist and I am trying to make sense of the existing theories in this field from the viewpoint of a layman such as I am. I will therefore avoid using mathematics and try to make sense of it in my own specific way, which was one of my motivations to write this book.

To go back to the main topic, Einstein's theory of relativity states that gravitational energy does not only depend on mass density but also on pressure exerted outwards.

In other words, pressure exerted outwards increases gravitational force, which is pulling unwoods. So force of gravity is equal to mass density + pressure in all 3 directions as we have 3 dimensions of space. Therefore, pressure is multiplied with 3. As gravity opposes the expansion of the universe, it is negative in the relation of the expansion. It could therefore be expressed roughly with an equation such as:

– (mass density + pressure times 3)

The negative sign is due to the negative effect gravitation has on the rate of expansion, making it slow down.

We can even make some analogy between building of the confidence of a person and expansion of the universe. The building of confidence will be like expanding or increasing the strength of personality where positive thoughts about himself or herself would make the confidence grow bigger. Opposing thoughts (like gravity) or comments on someone's capability have a negative effect on his or her confidence and is therefore negative in relation to the building of their confidence. An example is a negative comment about someone's capability like: "You are not able to do it" or "It is not possible for you to do it in such a short time." However, if these persons are receiving a negative comment to a negative comment then the final comment becomes positive which will boost a person's confidence,making it become bigger. For example: "It is not that you are not capable of doing it" or "It is not true that you are not capable of doing it" or "It is not true that you cannot do it in such a short time."

Similarly, if pressure in the equation outlined above turns out to be negative, then negative in brackets with negative outside brackets gives a final positive value, making the universe speed up in its expansion instead of slowing down.

A possible source of negative pressure is looks to be in quantum fluctuation where a virtual particle is constantly created from an empty space and where there is temporary violation of the law of conservation of energy but quickly comes to zero net of energy following annihilation of matter and antimatter particles.

The energy for the creation of a virtual particle is borrowed from a vacuum. That consequently creates negative pressure in the vacuum. If particles do not annihilate each other to establish zero net energy of vacuum again, then the negative pressure in the universe will continue to exist which will be the driving force of universe expansion. This will happen if, for some reason, created partials and antiparticles do not meet each other so they continue to exist instead of being annihilated.

Let me try to explain this in a slightly different way. We can imagine an empty room. The space of this room is occupied with plenty of invisible boxes. Every now and then, two of the invisible boxes come out of the room, becoming visible. One of these would be matter and another, antimatter or virtual particles. Their place is now empty. In other words, we have negative pressure created as boxes are out, becoming visible or virtual particles. As boxes quickly fuse together, annihilating their existence, they go back to where they came from, becoming again invisible. In that way, negative pressure created is restored as the boxes are back. However, if the boxes continue to exist and do not annihilate each other, then their place in the room remains empty. We now have negative pressure which gives negative value when summarised with mass density in brackets. Negative value multiplied with negative value outside the brackets of formula gives positive value. It means that gravitation is now positive and enhances the extension of the universe.

An example:

– (mass density is say number 10 + pressure 10 x 3)

which is:

– (10 + 30)

which is:

– (40) which is finally – 40.

It is because if we multiply negative value (-) outside the brackets with positive value (+) in the brackets, the final result we get will be negative value (-)

If, however, we multiply negative value outside the bracket (–) with negative value *inside the brackets* then we finally get positive value.

– (mass density 10 + negative pressure of –10 x 3)

which is:

– (10-30)

which is:

– (–20) which is finally 20 and not –20

Quintessence is another theory or attempt to explain the reason behind negative energy. Unlike the cosmological constant, which does not change in space and time, quintessence, as a form of dark energy, will vary in space and time. In the *Encyclopedia Britannica* it is defined as *a transient vacuum energy resulting from potential energy of a dynamic field.*

Quintessence is referred to as a fifth element. In the time of Aristotle it was believed that all matter on Earth was composed of 4 elements: earth, water, fire and air. Aristotle believed that all celestial bodies and the space above Earth, heaven, were composed of ether or from the fifth essence, which comes to be quintessence from the Latin.

Some scientists today regard dark energy as a fifth fundamental force.

Unlike the cosmological constant which is based on the logic of negative pressure in empty space, quintessence tried to explain the existence of dark energy in different ways. **Professor Alex Filippenko** was giving an example about broken symmetry in one of the documentaries about dark energy, which is easily accessible on YouTube. Instead of me trying to paraphrase what Professor Filippenko was teaching us, I would recommend watching this documentary.

What we know today is that the expansion of the universe is accelerating. The reason behind it is still part of speculation with as yet no theory emerging to be proven or confirmed by experiment. Therefore, there remain still a number of possibilities to explain the current observation of the universe, including one where the concept we have today about gravitation based on the theory of relativity can be changed.

6
SPACE-TIME

Curved space-time tells mass-energy how to move;
mass-energy tells space-time how to curve.
John Wheeler

John Wheeler, physicist, was the first American involved in the theoretical development of the atomic bomb. He worked with **Niels Bohr** in explaining the principle in relation to nuclear fission. He also gave ideas for names such as *black hole* and *wormhole.*

His quote about gravity and relation to space-time contains basically all that is needed to understand or have a rough idea of this concept.

Mass curves space-time fabric through time dilatation and space contraction, while curved space-time fabric accelerates movement of the mass towards the massive object which in the first place creates curve of space-time, making the other object fall towards the massive object in a circular movement or rotation which becomes faster and faster as it approaches the massive object.

Time and space are physical characteristics or property of the universe. The space between galaxies is

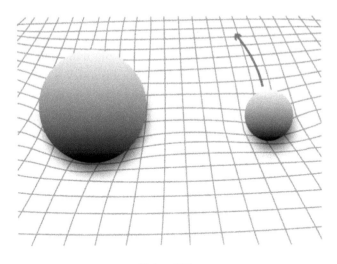

Picture 6.01

empty space but it is space nevertheless. When we think about the idea of the creation of the universe as some initial hot dense place less than the size of an atom, which explodes to become a universe, expanding ever since then, the image that is immediately created in our mind is a scenario where this explosion occurs in an empty space; the space, which we know, and which can be imagined that exists as such between galaxies or between any celestial bodies in the universe. This is not what really happened at the explosion of the universe.

The universe was created in a big explosion as the Big Bang theory explains, but this was an explosion of energy and matter as well as of space and time. There was no time and there was no space before the Big Bang happened. The beginning of the universe starts from zero time and zero space. To go into the philosophy as to why there was nothing before the Big Bang and why the universe developed into the way it is today is beyond my capability as an ordinary layman. In *Foundations of Modern Cosmology* by John F. Hawley and Katherine A. Holcomb it is stated that: *Because science must deal with physical entities, the issue of the creation of the universe is necessarily metaphysical.*

When fundamental constants of nature are considered such as speed of light, Planck's constant, charge of the electron and the gravitational constant, to mention just a few, it can be said that they have an exact value needed to be there to create matter in the way it is created which creates conditions to support the creation and existence of life within the universe. One of the ways which can be used to explain why the universe has developed in this way with such particular characteristics is the so-called *Anthropic_Principle.* According to this principle, it is a presence of life which determines the development of the universe in such a way that permits creation and existence of life.

It was **Fred Hoyle**, British mathematician and astronomer, who used anthropic reasoning, predicting that the intrinsic energy level of a carbon atom has almost equal value as the sum of the energetic value of 3 helium atoms in the core of a star to allow fusion of these atoms of helium into a carbon atom to take place. Such nuclear reaction progresses under these conditions very quickly. It is because it is said to be *resonant.* There was, therefore, the exact condition created within a star for a carbon atom to develop which will serve later on as a building block of organic compounds leading to the creation of life. Every living organism is composed of chemical compounds and each organic chemical compound has carbon atoms as a core of their structure.

It was found out that the next chain in the nuclear reaction involving the fusion of one more helium atom with a carbon atom to create an oxygen atom failed to be resonant and, as such, prevented carbon atoms burning to oxygen. It was due to the different energy levels of oxygen, carbon and helium that this nuclear process went in such a way within the star that carbon was created and remained safe. Combinations of the value of the fundamental constants of nature determine such energy levels. If the value of constants is slightly different then the outcome of nuclear reaction will be different with no possibility of creating carbon atoms in abundance.

Carbon atoms to be released from a star require the star to die in a supernova explosion. In this process, many elements with higher atomic numbers were made. The average lifespan of a star is billions of years. As the universe expanded through this time, it was necessary for the universe to be very big in a size of billions of light years before conditions were met such that planets were made of dust which was left as a remnant of a supernova explosion. Such cloud contained the elements to build rocky planets and a number of necessary elements for life to be created.

This is one way of looking at the question as to why the universe developed the way it did.

I should go into the nature of space and time as a property of the universe, but before that, I believe it is the right place to mention a few things about the size and shape of the universe.

We are able to observe the edge of the universe, which corresponds to a time around 380 000 light years after the Big Bang when photons decoupled from matter and went away from the matter in every direction. These photons are now reaching us from all directions, not in the spectrum of visible light but in the spectrum of microwave background radiation. These signals of cosmic microwave background (CMB) travelled to us from a distance of around 13.7 billion light years, which is a snapshot of that time of the universe when photons decoupled from matter. That corresponds roughly to the beginning of the universe which is the reason for the statement that the universe is 13.7 billion years old. It also corresponds to the edge of the universe at that time, 13.7 million years ago. This is, however, not the current edge of the universe. (I am referring here to the edge of the universe or size, not the age of the universe or how old the universe is.) The reason it is not the current edge of the universe is that the universe has expanded from the time we received CMB radiation which was the edge of the universe 13.7 billion years ago when these photons left these areas and have now reached us after 13.7 billion years. Meanwhile, the universe has expanded and it is estimated that now its radius is around 46 billion light years or diameter around 92 billion light years. These calculations are made taking into account the Hubble constant that refers to the rate of the expansion of the universe. It is how I am making sense of it.

WMAP, which stands for the Wilkinson Microwave Anisotropy Probe, is a spacecraft (also called Explorer 80) launched by NASA in 2001, which is equipped with technology to measure differences in the temperature within CMB radiation called anisotropy of CMB radiation. The data collected supports the current standard theory or standard model of cosmology called the *Lambda-CDM model* which is modified Big Bang theory to take into account cosmological constant (lambda) and cold dark matter (CDM). This theory is more precise or reflects more accurately or gives a more plausible explanation of the origin of the universe based on the Big Bang theory.

WMAP confirmed that the universe is flat in shape in 2013 with accuracy with only 0.4% of margin error. (Details taken from NASA website: http://map.gsfc. nasa.gov/universe/uni_shape.html)

It is consistent with the Big Bang theory and the inflationary theory. The reason that the universe is in flat shape is ratio between mass density and critical density which is almost equal to 1.

What does it mean?

We know that gravitation depends on mass density and pressure. There are a minimum amount of particles which need to be per unit of volume for gravitational force to remain and exert or have an effect in this space.

This can be called critical density of space. If this space has more mass density than critical density then we can expect that this space can expand to the maximum point but then will start to contract, as mass density is bigger than critical. How far this place will extend before it starts to contract will depend on how much higher mass density is than critical density. If the universe has a very higher mass density than critical density then it would not be able to expand more than is the size of our Milky Way galaxy or less, again depending on mass density over the critical density. If the mass density is less than the critical density, the universe will go on expanding indefinitely.

When mass density is equal to critical density then the universe has a flat shape.

The following analogy to understand it better is not the best I can think of as it can be challenged very easily, but I will use it anyway.

When we pour water into a pan it does not take the shape of a ball but fills the pan to a depth of 1-2 ml (if we pour just a little bit of water). We can say that water lies in a pan and has a flat shape as it does not have enough mass density which will pull molecules water together to make a ball shape. If we now add to this flat-shaped amount of water, flour or chocolate powder, then the density of a mass will increase above the critical density, making the powder and the molecules of water come together. Now we are able to make a ball-shaped mixture of water and chocolates.

In the previous chapter, dark energy is explained as

acceleration of expansion of intergalactic space. A galaxy on its own does not expand within itself in the sense of getting increased distances between stars in galaxies. That is because the mass density is bigger than the critical density within the diameter of galaxies (there is stellar dust, for instance), which make them, overcoming the effect of dark energy so it is only that intergalactic space is increasing.

I cannot resist mentioning the importance of the current scientific approach to studying nature, taking into account all needed to arrive at a theory or theories which will give a logical and accurate explanation of phenomena we observe in our universe.

The basic approach is the **Copernican principle**. It is based on the principle that if a theory requires a special condition around which the theory needs to bend, then this theory cannot be plausible or reflect what is really happening in relation to a particular phenomenon which the theory tried to explain.

An example is the Ptolemaic geocentric system, which puts the Earth at the centre of the universe. In order to keep this theory alive or plausible, Ptolemy created the theory of epicycles, which explained the movement of planets then known. So binding the theory around the presumption that the Earth is the centre of the universe, he created the theory, which looked plausible but could not exist long, as it did not reflect the reality.

We know that Earth is not the centre, but goes around the Sun. We know that the Sun is not the centre of the universe and it goes around the Milky Way, our own galaxy. Our own galaxy is not the centre either and is only one of many galaxies in a local cluster of galaxies.

On a large scale, the universe is isotropic and homogeneous.

Isotropy refers to a situation where there is no special direction wherever you are placed in the universe. The universe would look the same in any direction, regardless of the place we are in the universe. This does not appear to be at small distances for the universe. For example, if we look locally within our solar system then of course directions look different as we have our sun giving us daylight, the Moon that looks bigger than other celestial bodies. Stars within our galaxies have particular positions creating particular shapes in the sky called constellations. However, when the universe is observed on a large scale with distances of millions of light years then it does look the same in any direction. In other words, the universe is isotropic.

Homogeneity means that every place in the universe is the same and not different to the others. The fact that the universe is homogeneous is important as thanks to this we are able to detect the same elements such as helium or hydrogen that exist everywhere in the universe, regardless of the place we pick to observe.

The universe is isotopic and homogeneous which means there is no centre. Because there is no centre then there is not a particular place where the Big Bang occurred. In other words, the Big Bang has occurred everywhere. Every point in the universe is the centre of the universe where the Big Bang occurred.

The best way to explain it is if we imagine a birthday balloon before it is inflated. This balloon has its particular shape and is composed of molecules or atoms, which make its wall. Let's imagine these atoms or molecules as huge, almost indefinite numbers of dots tightly situated next to each other, making the wall of this balloon. Now we start to inflate this balloon. As the surface of the balloon enlarges, the distance between each dot enlarges as well. If I am at one of these dots (being reduced in my size by some technology) then I will see how all the other dots are moving away from me in all directions. That would be the same experience if I were at any other dot. I would be able to see other dots moving away from me in all directions until they move away from me beyond a distance of 13.7 billion light years. Afterwards, I would not be able to see them any more as it will take time for light from those dots to come to me. 13.7 billion light years is the diameter of the observable universe in all directions and is the same from whatever position we are looking. It would be the same if we look at the universe from our planet or from a planet that is located in the Andromeda galaxy, for example.

Momentum Conservation Principle

In Chapters 1 and 3 I gave an explanation of the conservation of matter and energy in the way I could make sense of it. Einstein's famous formula, which reflects the possibility of energy change to a matter and vice versa, requires modification of conservation of matter and energy as a separate entity into a more precise statement of conservation of mass-energy within a closed system or the universe.

There is one more very important law of conservation, which is referred to as the **conservation of the momentum**.

The conservation of the momentum is simply defined as a *product of mass and velocity*. As it refers to velocity or movement of the mass in any direction of tridimensional space, I found that this chapter is the right place to mention a few things about this law.

Conservation of momentum means that in a closed system where there are no external forces acting on a system, the total momentum of the system does not change.

When I tried to make sense of the conservation of energy or internal energy within the closed system I outlined that the internal energy can change if energy from outside is transferred in the system. This transfer will take place by the help of work or heat. However, if the system is closed and there is no outside from where energy can be inserted to the system, then the internal energy within the system remains conserved, unchanged.

Like internal energy is changed if it is inserted in the system by work or heat, so the momentum is changed if velocity changes, which means that force is exerted from outside.

To understand it better, it is important to keep in mind that momentum is product of mass and velocity:

momentum = mass x velocity

Velocity is a change in the position of the object or its displacement, which takes place over a period of time. It has a vector property as it applies to a particular direction.

Force is, however, described as a product of mass and acceleration:

force = mass x acceleration

Acceleration is a change of velocity over a period of time. So the above equation can be arranged as:

$$\text{force} = \text{mass x } \frac{\text{change of velocity}}{\text{over a period of time}}$$

Mass x velocity is momentum so we can replace the right side of the above equation with momentum. However, the velocity changes over a period of time. As momentum is a product of mass and velocity and velocity is changing over a period of time, then product or momentum changes over a period of time as well. Therefore, force is a change of momentum over a period of time:

$$\text{Force} = \frac{\text{change of momentum}}{\text{over period of time}}$$

Ultimately, we have a change of momentum if external force is exerted on the system from outside. If the system is closed with no external forces, then momentum before motion and after motion within the closed system remains the same.

If two objects are on a collision path within the closed system then their total momentum before collision is equal to the total momentum after collision. When we hit a tennis ball with a racket then the ball will fly away but the product of ball mass and velocity of the ball will be equal to the product of velocity by which the ball was hit and the mass of the racket.

racket x velocity = tennis ball x velocity

The ball velocity will be higher than racket one but the mass of the ball will be less than racket one. Therefore, the product of the corresponding masses and their velocity will be the same. In this particular case, the racket will lose its momentum upon collision with the ball while the ball will gain the same momentum.

There are two types of momentum:

Linear momentum or translational momentum. This is the product of mass and velocity of an object, which is in motion linearly as explained above.

Angular momentum or rotational momentum is the product of moment of rotational inertia and rotational velocity.

Angular momentum = moment of inertia x rotational velocity

Moment of inertia is rotational analog to mass in linear momentum as rotational velocity is to velocity of linear momentum. Rotational velocity is angular velocity, which can be mathematically described as ratio of velocity and radius.

As with linear momentum, angular momentum is conserved within the closed system as long as it is not affected by applying external forces, which in the case of angular momentum is a torque. Torque is influence, which tends to change the rotational motion of an object. Angular momentum is very important in astronomy.

In the history of Earth, the prevailing theory is that another planet hit our planet at an early stage of her life. The result of this collision was displacement of the Earth's axis, which tilts at around a 23-degree angle. It is thanks to this that we have seasons. The other thing which happened is that in this collision the Moon was formed from debris and was much closer to the Earth than it is today. Due to conservation of angular momentum, Earth rotated faster when the Moon was closer to Earth. The same applied to moon velocity being faster around the Earth at that time. As the Moon moved away, its orbital velocity as much as rotational velocity of the Earth reduced. A day lasts twenty-four hours in present time but was much less than twenty-four hours in the distant past of Earth's history when the Moon was closer and Earth consequently rotated faster. This change of rotation with the change of the distance of the Moon from the Earth is due to conservation of angular momentum. To explain this I have to introduce a little bit of mathematics.

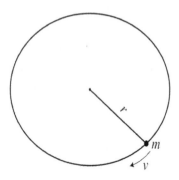

Angular momentum=m·v·r

Picture 6.02

Angular momentum is actually the product of mass (m), velocity (v) and radius (r) from the centre.

Angular momentum is conserved and not changing. Therefore, the product or m, v and r is constant.

Velocity comes from angular velocity. Angular velocity is ratio of velocity (v) /radius (r). From there, we can work out that velocity is the product of angular velocity and radius. When we replace v in mvr product with angular velocity times r, we get a final equation for angular momentum which is:

mass x angular velocity x radius on squared

This result is constant, does not change as it represents conservation of angular momentum.

Angular momentum = m v r

Therefore: m v r = constant

Velocity (v) is coming from angular velocity (ω)
Angular velocity is:

$$\omega = v \text{ (velocity)} / r \text{ (radius)}$$

From this equation, velocity is:

$$v = \omega r$$

It is the same as:

$$3(\omega)=6(v)/2(r) \text{ where}$$
$$6(v)= 3(\omega) \text{ times} 2(r)$$

So in angular momentum, v can be replaced by ωr:

Therefore, angular momentum = m v r is:

Angular momentum= m ωr r

(r multiplied by r = r^2)

So the final equation for angular momentum is:

angular momentum = m ω r^2

m = mass

ω = angular velocity

r^2 = radius

This means that if the distance or radius from the centre

is reduced, then the angular velocity or speed of spinning has to increase for the product to remain constant. Equally, spinning or angular velocity will be reduced if the radius from the centre is increased, as the product has to remain constant because angular momentum is conserved. So when the Moon was closer to the Earth, radius or distance of the Moon to the centre of Earth was smaller. Consequently, the Moon was spinning around the Earth faster and with it, Earth rotating faster. Here, there is also another effect which the Moon exerted to the Earth, and that is the gravitational effect on the oceans causing tides. I will not go into this now.

An ice dancer spins faster with her hands close to her chest. With her hands and legs outstretched,she would spin slower

6.03

Picture 6.03

The illustration above is just my attempt to demonstrate that an ice dancer spins faster when her hands are brought together close to her body (chest). If she spins with her arm and legs outstretched, then she is not able to spin so fast.

In modern science and physics in particular there is one very important entity, which is tightly related to laws of conservation. It is called invariance.

Invariance is a property, or quantity, which does not change and is always measured having exactly the same value regardless of the conditions under which measurement is taking place. Speed of light is an example of invariance. Conservation of baryonic or lepton numbers is another example.

If the closed system, with the internal energy of a certain value, gets larger or space increases within the system, the quantity of internal energy will remain the same. Roughly speaking, the number of energy units per unit of volume will be less in increased space compared to the time when the system was reduced. However, the total amount of internal energy will not change with a change of space and time. This is obviously when we are considering a closed system with no surroundings. The same applies to linear and angular momentum.

We know now that speed of light is constant and does not change regardless of the conditions under which the speed is measured. It is therefore an example of invariance.

Before establishing this as a fact, it was known that everything is in motion within the universe and whether some particular object moves or not depends on the relation this particular object or matter has with another object, which is not in this same frame of motion.

The situation where an object moves with constant velocity is called an **inertial reference frame**. This also refers to a situation when an object is at rest. An inertial reference frame refers to these two occasions where we do not have acceleration. In any other situation, where an object is moving so that its velocity changes and acceleration or deceleration is present, it is not an inertial reference frame. It is called **non-inertial reference**.

When we are in an inertial reference frame (with a constant speed) we are not able to know if we are moving or not unless we compare our reference frame with another. If ten people are in the same reference frame then there is no way they will know or they can test whether they are moving or at rest if they focus their attention on physical phenomena taking place in this same reference frame. The reason is that all objects move with the same speed in this inertial reference frame and therefore look as if they are at rest within this inertial frame of reference. If we are sitting in a train with another passenger in front of us and are moving at a speed of 20 miles per hour, we are not going to reach a passenger in front of us due to this speed. The distance between us and that passenger will remain the same because he moves at the same speed as he is in the same reference frame. We can also easily pour coffee into a cup without spilling it all over as all of these are in the same reference frame, moving at the same speed. It looks as if we do not move.

There are two ways we can test and confirm that we are moving. One is to look outside and see trees moving away, for instance. The other way of finding out that we are moving is if the train accelerates or decelerates or changes its velocity. With the change of velocity, we are no longer in an inertial frame as explained above.

The fact that we cannot know if we are moving when we pay attention to events taking place in our reference frame was pointed out by Galileo. It is also called Galilean relativity.

Before going further, I would like to emphasise the difference between velocity and acceleration.

Velocity is the change of position of an object from place A to place B over a period of time. This change takes place in a particular direction such as left, right, up and down or backwards and forwards. Therefore, velocity has vector property and that characteristic is very important in defying velocity. The magnitude of velocity is measured by speed.

Acceleration is a change of velocity.

Mass multiplied with acceleration (change of velocity) is a force.

force = mass x acceleration

Acceleration is a change of velocity so

force = mass x change of velocity

Whenever there is a change of velocity we have a force applied on a particular motion, which was at a constant velocity (inertial reference frame) or at rest before force changed it. That is in agreement with Newton's first and second laws of motion which state:

First law: an object is either at rest or moves at a constant velocity unless acted on by external forces.

Second law: it basically refers to a change of velocity or change of momentum when an object is subject to external forces.

We have a change of velocity when velocity:

- Increases
- Decreases
- Changes direction

An increase or decrease in speed is obviously due to force being applied to a particular motion, which is at constant velocity or an inertial reference frame before the force was applied.

However, when the direction of constant velocity changes then we have acceleration even if we have a constant speed. An example is orbital movement or circular movement with a constant speed of 20 miles per hour all the time around. Yet we have change of velocity or acceleration, which we can feel very much.

How can we have acceleration there when speed is constant?

Change of speed can be another way of describing acceleration.

So how can we have acceleration when we have a constant speed?

The answer is in the fact that velocity is a vector and refers to a specific direction.

If an object travels at a constant speed of 20 miles per hour in the direction of left to right then it remains in an inertial reference frame with constant velocity only for this direction. The constant velocity of this object in the direction of up to down is zero while it is in an inertial reference frame of moving left to right. If this object now changes direction of movement from left to right to direction from up to down, remaining at the same speed, the acceleration will be noticed as the object suddenly moves in a direction which until that moment has zero velocity. So, although the speed does not change, the acceleration will happen upon entering a new direction with zero speed for that direction before entering it. Therefore, in any circular motion there is acceleration, which is very much noticeable in everyday life, when we drive a car or motorbike on a bendy road, for example. As change of velocity or acceleration implies force then we have presence of force, which really is not exerted from outside. This force is experienced when an object is moving in a circular motion although it is not a force applied from outside and is only the result of acceleration due to change of velocity as the consequence of a change of direction (not taking here into account gravitational force). That is why centrifugal force is referred to as a fictitious force.

Any other motion, which is not at a constant velocity, is referred as a non-inertial reference frame. Rotational movement of the object is one of these frames as velocity constantly changes due to constant change of direction.

Before it was known that speed of light is a constant, it was believed that time and space can be regarded as invariance, or the separation in the time between two events or separation in a space between two places is a constant and not changeable. Speed, including the speed of light, was considered to be relative. Speed, i.e. how fast an object is moving, depends on the reference frame from where the measurement of this speed is taken.

If we are on a train moving at 80 miles per hour and throw the ball in the direction of the train's movement at a speed of 5 miles per hour, we will measure this speed to be 5 miles/hour as we, and the ball, are in the same reference frame. Our inertial reference frame is 80 miles per hour.

However, for a man who is standing outside the train and is in a rest reference frame compared with our reference frame, the ball's speed will be 85 miles per hour, which is the sum of the speed of the train and ball moving in the same direction. If the ball is thrown in the opposite direction of the train's movement then the ball's speed for the man outside the train will be 75 miles per hour. For the man on the train it will remain the same or at a speed of 5 miles per hour. This is logical and does not require more detailed explanation. These calculations are needed when we relate two different frames. They are called **Galilean transformations**. It refers to Galilean relativity and time when space and time were regarded as a constant and unchangeable value. It was Galilean relativity where also the speed of light was regarded to be such in relation to so-called ether. Basically, it was believed that the speed of light moves at such a speed only in relation to ether or another reference frame.

Galilean transformations are still valid and applicable to reference frames where speed is not significant but become less reliable with increased speed particularly going towards the speed of light. In this case, it is **Lorentz transformations** that are precise and relevant.

Once more, clearly to outline that observations which are done within the same inertial frame of reference cannot prove that we are moving, as all objects in this frame of reference move at the same constant speed. This is referred to as Galilean relativity.

Once more, to repeat that close to the end of the 19th century, it was believed that light moves through space at such a speed only in relation to ether or another reference frame.

Having in mind Galilean relativity and the possibility that light moves at such a speed according to ether, Einstein did a so-called mind experiment when he was only sixteen years old in 1886.

I will outline this experiment in a way which helps me to understand it.

Light is composed of photons, which can be regarded as light particles. We can imagine a train travelling at a speed of 300 000 kilometres per second and that each passenger on the train is a photon. The train is a beam of light, which has photons (passengers) travelling at a speed of 300 000 kilometres per second.

We have to keep in mind that the train (beam of light) moves at the speed of light according to another reference frame. That is a train platform and people on it. All of these make *ether* according to which the light beam (train) moves at the speed of 300 000 kilometres per second.

Now let's imagine a girl is standing on the platform looking in a mirror. She is in another reference frame to the train (light beam). She is part of ether. The photons will leave her face at 300 000 kilometres per second and bounce from the mirror towards her eyes at the same speed. She will see her reflection in the mirror instantly due to the light speed.

Now we can imagine that our girl has got on the train and is now moving at the same speed of 300 000 kilometres per second. She is not any more part of ether and is moving at an exact speed like photons or people on the train in relation to ether (platform station and people standing on the platform, the place where she was before getting on the train). Now when she sits behind a passenger, she cannot reach him, as he is moving at the same speed forward as she is. In other words, if she now looks in the mirror she would not be able to see herself in the mirror, as photons cannot move from her face to the mirror because both (her and photons on her face) travel at the same speed.

Einstein did exactly this experiment, imagining what would happen if he travelled at the speed of light and with a mirror in front of him. He would then travel at the speed of light in relation to the ether as light does. Light, therefore, cannot leave his face to reach the mirror and come back due to the fact that both him and lights or photons travel at the same speed in relation to ether or in relation to other frames of reference. This is, therefore, violation of the principle described by Galilean relativity, as it is now possible to know that he is travelling at the speed of light, looking at the events happening in his own inertial reference frame. That is because he would not be able to see himself in the mirror as he is travelling at the same speed as light or photons.

What would happen if photons did not move with such speed relative to other frames of reference?

What if they moved at such a speed relative to an observer regardless of the frame of reference the observer is in?

This would mean that photons leave the girl's face towards the mirror and bounce back to her eyes at the speed of 300 000 km per second when she is at rest in the train station as well as on the train at a speed of 300 000 kilometres per second. It means that the speed of light moves at such a constant speed in any frame reference an observer is in. In such a situation, the speed of light does not change and remains the same in any reference frame and is, therefore, an example of invariance.

In this situation, however, we cannot apply Galilean relativity and Galilean transformations.

In the train and ball example, the speed of the ball was different for a man standing in a train speeding at 80 miles per hour (the ball for that man was travelling at a speed of 5 miles per hour) and the man standing on the ground. The man on the ground at rest measured a speed of 85 or 75 miles per hour depending whether the ball was thrown in the direction or opposite direction of the train's movement.

The girl on the train moving at light speed can now see her face as light from her face moves in relation to her with the same speed to the mirror and back. But if we apply Galilean transformations and Galilean relativity then the man standing on the ground will see photons leaving her face at a speed of 600 000 kilometres per second as she herself is moving at a speed of 300 000 kilometres per second. It would be logical of Galilean relativity and transformations to make a sum of the speed of train 300 000 km/s and speed of photons leaving her face 300 000 km/s to get a speed of 600 000 km/s.

That is not possible because light is perceived from the man's reference frame (at rest) as also travelling at a speed of 300 000 km/s as it is for the girl's reference frame on the train. The only way the man can perceive her seeing herself in the mirror is that with such speed, the distance between her and the mirror has reduced in the direction of the movement of the train. That is only perceived from his viewpoint or reference frame at rest. In this case, the distance between the mirror and the girl's face has reduced to zero. The time passing by for photons leaving the girl's face has also reduced to zero for the man who is at a rest reference frame in relation to the train. That means that we have space contraction and time dilatation to keep constant speed of light to be seen as such from any reference frame and only in relation to an observer.

To put it another way, light travels at the same speed of around 300 000 km per second for the girl on the train and the man on the platform. For the girl, photons leave her face, hit the surface of the mirror and bounce back to her eyes so she can see her reflection. For the man on the platform, photons reach the mirror as well and go back to her eyes. However, such things are possible only because her mirror is now fused with her face for the perspective of the man on the platform (distance is contracted to zero at the speed of light). Also, time has stopped at the speed of light from the point of view where the man is on the platform so photons cannot leave the girl's face but photons again reach the mirror as the mirror is now fused with the girl's face. Therefore, the speed of light remains constant at around 300 000 km per second for the man on the platform as well. It is because the girl is moving at that speed.

We therefore here apply the Lorentz transformation and Einstein's theory of relativity.

In 1887, **Albert Michelson** and **Edward W. Morley** set up an experiment to test or detect ether to confirm that light moves with such speed due to ether. The philosophy behind the experiment was that light would have a higher speed if both the Earth and light moved in the same direction in relation to ether (other reference frame) or a lower speed if light moved in the opposite direction to Earth's movement in relation to ether. The same principle applied to the train and ball-throwing in the train as explained in that example previously. The test confirmed that the speed of light remains the same and that ether does not exist. It confirmed that light travels at the same speed irrespective of whether the source of light is moving or the place from where it is measured is moving.

All the progress made in science and physics to that moment, and particularly tests confirming the constant speed of light, set the scene for Einstein to make a drastic change in how space and time is perceived. He developed his theory of special and general relativity.

The Special Theory of Relativity was published in 1905. This theory is basically related to changes taking place in different inertial reference frames in relation to each other. It does not involve acceleration. Therefore, it applies only to movement with constant velocity in a straight line as any change of direction (bending or circular movement) would imply acceleration and would not be an inertial reference frame. Obviously, standing or positioned at rest is also an inertial reference frame.

The General Theory of Relativity deals with non-inertial

reference frames or with cases where acceleration is involved. It actually explains gravitation to be caused by curvature of a space-time made by matter, which has a large mass density. The larger mass density and pressure exerts the larger curvature of space-time.

Both theories are basically referring to time and space, which are not constant. Their quantities are different to two observers who do that observation from different frames of reference to each other. How different these quantities will be depends on how large the difference in speed is between two reference frames.

THE SPECIAL THEORY OF RELATIVITY

Einstein published The Special Theory of Relativity in 1905. His theory is based on two postulates. They are:

1. *The laws of nature are the same in all inertial frames of reference*
2. *The speed of light in a vacuum is the same in all inertial frames of reference*

Before my attempt to explain this theory in my way, I would like to start with one short story.

Alison was preparing to go to a business meeting at a business centre, which she needed to attend on that day at 2pm. She arrived at a luxury 5 star hotel near the business centre the day before. She was given instructions for the exact location of the business centre. She needed to go forward to the end of the high street (around 400 metres) and then turn right for another 200 metres where she would reach the business centre. She needed to go by lift up to the third floor where the meeting would be taking place at 2pm.

What Alison was given were coordinates of a precise location in a space, consisting of three dimensions. She needed to be there that day at 2pm, which is another dimension or time dimension. Four dimensions, three space dimensions and one time dimension therefore determine the location of the object. We would not be able to allocate Alison at the business centre at 11am, as she would not be there before 2pm.

Alison left her hotel at 1.30pm and walked at a steady speed, reaching the business centre at exactly 2pm. She was walking through three dimensions of space and dimensions of time with the constant velocity at a speed of 1200 metres per hour. She was able to pass 600 metres in half an hour, from her hotel to the business centre. The trajectory of her movement can be traced through time and space if we make a diagram. It is very difficult to draw a diagram of space outlining· all three dimensions and particularly if we need to add another dimension: time dimension. The diagram can be simplified by using only one dimension of space, say only from left to right or right to left, and time dimension.

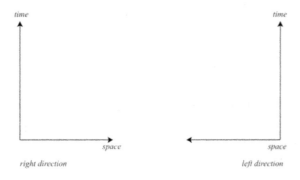

Picture 6.04

Left to right space dimension is outlined with a horizontal axis while time dimension is outlined with a vertical axis as shown in the illustration above.

We can make diagrams of movements in space and time of many different objects that move at different speeds. In other words, we can make and compare different reference frames in a time space diagram. However, in order to be able to do so, we have to have the same unit for time and space. We can do that by using light year, hour or second for distances which the light passes in a year, hour or second. We can do that because light has a constant speed and therefore there is an exact distance it will pass in a particular unit of time. In the diagram of space-time below, I used light seconds for measuring distance in space. Each light second actually represents a distance of 300 000 km which is the distance the light passes in a second. If we now plot the location the photon has after 1 second passing a distance of 1 light second and continue to do so, then we can draw a line through these locations the photon was at each second and the distances passed during this second. The drawn line is the trajectory of photon movement through space and time at the speed of light. This line is at 45 degrees in relation to space dimension and it is important as it represents the speed of light.

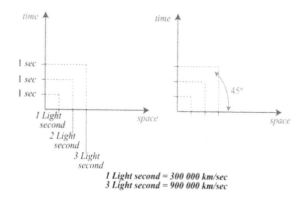

Picture 6.05

Using this kind of diagram, nothing can travel with a trajectory of less than 45 degrees as this would imply speed, which is faster than the speed of light (No. 1 diagram in illustration below). In this case, this object will pass larger distance for a shorter time than light does and this is not possible as there is nothing faster than the speed of light. This is therefore not allowed. Also, moving in the opposite direction (No. 3 in illustration) is not allowed as this would imply moving backwards in time or going to the past and that is not possible.

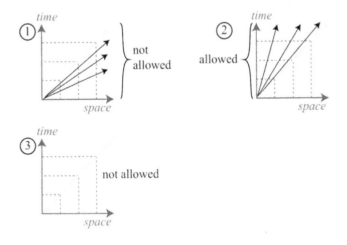

Picture 6.06

If we measure the speed and movement of other objects in such a time-space diagram, we have to use the light year as a space measure. As light is the fastest that anything can travel through space and time, then the speed of another object is or can be measured as a fraction of light speed. For example, if we said that an object travels at 1/10 of the speed of light that means that it takes 10 years for this object to pass the distance the light passes in 1 year. If an object moves at 1/5 of the speed of light it will takes 5 years for this object to pass the distance the light passes in 1 year. If the object travels at 3/5 of the speed of light it will take the object 5 years to pass the distance the light passes in 3 years.

I will now move to special relativity, which applies only to objects that are moving in inertial reference frames (they are moving either with constant velocity or are at rest).

Let's imagine that we have two men, each encapsulated in a closed container with a glass window so that they can see each other (like in two spaceships facing each other). If these two containers are in a space so they cannot see stars, Earth or are not able to see any other objects apart from each other and both containers are in inertial reference frames then they would not be able to detect which one of them is moving (this obviously applies if one of them is in an inertial reference frame applying to a constant velocity). In such a scenario, neither of them will be able to know which one is moving or whether their movement is the result of the movement of both of them. This is the focus of the special theory of relativity from where we have also what is called the twin paradox.

Basically, we have the change of the quantity of the time and length for an event when those two men, who belong to two different reference frames in relation to each other, measure this event.

In my following example, I will try to explain it in a way that it is similarly explained in almost any book you read about it. Perhaps a slight difference will be that I will overemphasise bits, which make me confused when I read about it initially. In other words, I pay attention to those things which help me to make sense of or understand this topic better. I have used mathematics in this section but obsessively will try to explain calculations in each step supporting this by using numbers on some occasions to make it more understandable.

I will use an example which demonstrates how time slows down for a man travelling on a train as perceived by a man who is at rest and who watches a train passing.

A clock will be used as a beam of light which leaves the floor of the train and hits the ceiling of the train. We can imagine this to be a photon. The speed of light is constant. If the distance from the floor to the ceiling of the train is constant and unchangeable then the interval of time that passes from the moment the photon leaves the floor to the moment the photon reaches the ceiling is always the same. We have therefore constant rhythms with the same pace of time separating two events in time (the moment the photon leaves the floor to the moment it hits the ceiling). With the constant unchangeable rhythm we can measure time passing. If something produces a bang noise every 1minute and does not change its rhythm then after we hear a bang 10 times we know 10 minutes have passed.

To go back to our light and photon, we will have the same rhythm or time between the photon leaving the floor and hitting the ceiling as the time needed for the photon to bounce back from the ceiling and hit the floor from where it starts its journey. This time interval will always be the same for the man standing in the train.

Let's imagine the impossible for the sake of understanding this topic. Let's imagine that it takes 1 photon 1 second to travel from the floor to the ceiling in the train. Let's imagine that another photon does not leave the floor until the previous one reaches the ceiling and such a pattern goes on. We will have a constant pace of time where after 1 second every successive photon will hit the ceiling 1 second apart. After 10 such events we would know that 10 seconds in time have passed. That is time which has passed for the man standing in the train and for the man watching time passing in that train from the ground. What is very important to clarify here is that this statement of the same time passing for the man on the train and perception of this time passing in the train for the man on the ground applies only if both the man on the train and the man on the ground are at rest (in the same reference frame) as in the illustration below.

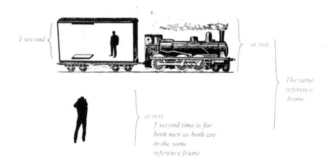

Picture 6.07

If the train moves at a high speed then we have a completely different scenario. For the man who stands in the train there will be no change in the time needed for the photon to reach the ceiling from the moment it leaves the floor. The time passing or clock ticking represented by photons is not changed for the man on the train as it is in the same reference frame with the clock (in this case, photon) which is on board the train. However, for the man on the ground the clock ticking on the now moving train has slowed down.

Why does this happen?

The reason is that for the man on the ground at the moment when a photon leaves the floor, the train moves a bit forward as it is not at rest any more. By the time the photon reaches the ceiling, the train will have moved so much that light or photons will travel diagonally in order to reach the ceiling. This distance still remains the same for the man on the train and goes up in a straight light, as both the photon and the man are moving at the same speed. Time has not slowed down for the man on the train as he has still the same rhythm or time passed for the photon to travel from the floor to the ceiling. However, for the man on the ground the distance from the floor to the ceiling in the train is not any more an upwards straight line. It is now diagonal which is a longer distance.

Now if we can have a situation where light does not have a constant speed and can adjust to a different situation than in this diagonal distance, the light can speed up to reach the ceiling at the same time as it does when it passes a shorter distance in the straight upwards line. In this situation again, both men (on the train and the ground) will measure the same time. However, **light has constant speed** and cannot speed up or slow down. Because of that, it will now take the photon a longer time to reach the ceiling, as the distance from the floor to the ceiling is longer. It means that we will have slower rhythm of that photon clock for the man standing on the ground. Although time did not change for the man on the train, the man on the ground perceived an increased time needed for the photon to reach the ceiling, which means an increase of the interval between two ticks and therefore he perceived that time has slowed down for the man on the train. This is only perceived from the position or observation made by the man who is standing on the ground.

In the illustration below I have tried to make it clear. I did not draw the other man on the train in order to make the picture clear, but imagine that the other man is on the train. The man on the train will not have a change of time as rhythm will not slow down, as for him the distance from the floor to the ceiling will remain a straight upwards distance. For the man on the ground, the rhythm will slow down as light now has to travel a longer distance, which will require a longer time to reach the ceiling. This will slow down rhythm which means that time will slow down for the man on the train as perceived by the man on the ground.

Picture 6.08. The blue cross on the diagonal line roughly corresponds to the distance a photon travels in a straight upwards line. The straight upwards line (not diagonal) is perceived by the man on the train. (The man on the train is not drawn here for the sake of getting a clearer picture.) So when we look now at the diagonal line which the photon has to travel to reach the ceiling (as perceived by the man on the ground), the distance, which corresponds to the straight up line, is up to the blue cross. The rest of the diagonal line is the additional distance that the photon needs to travel to reach the ceiling as perceived by the man on the ground.

We now have a triangle with sides where we can use Pythagoras' theorem to find out the size of the hypotenuse.

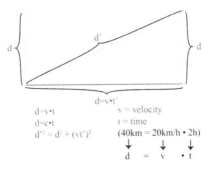

Picture 6.09

The hypotenuse is the distance from the floor to the ceiling or diagonal line the photon is travelling as perceived by the man on the ground. The initial distance from the floor to the ceiling perceived by the man on the train is a straight up line and can be marked as **d**. The diagonal line can be marked as **d'**. If we know the constant speed of a moving object then we can work out the distance if we multiply it with time:

$d = v \times t$ (velocity multiplied by time)
An example is that with a constant speed of 20 km/hour, in 2 hours we will pass 40 km.

Therefore, the distance the train will pass by the time a photon reaches the ceiling for the man standing on the ground is **v** times **t**, so we have defined all three sides of the triangle as **d**, **d'** and **v** times **t'**.

The velocity of this train is not known but the velocity of the photon is, which is the speed of light. So **d** is equal to **c** times **t**, while **c** times **t'** is equal to **d'**. So to find out d' can be formulated as:

$(ct')^2 = (ct)^2 + (vt')^2$

I forgot to mention that Pythagoras' theorem states that squared on hypotenuse is equal to the sum of squares of the two other sides of the triangle, which is in this case:

$d'^2 = d^2 + (vt')^2$ (look for this equation in the earlier illustration with the drawn triangle in it). Now we can go back to the last written equation, which can be written in such a way that we move **ct** squared from the right side to the left side of the equation:

$$(ct)^2 = (ct')^2 - (vt')^2$$

The logic behind it can easily be seen if we use numbers as an example:

$6 = 4 + 2$ so if we move 4 from the right side to the left side we get the equation where:

$4 = 6 - 2$

We can now write the above handwritten equation as:

$$c^2t^2 = c^2t'^2 - v^2t'^2$$

Again, it is mathematically correct and it can be checked by using numbers as an example:

$$(2 \cdot 3)^2 = (6)^2 = 36$$
$$\text{or}$$
$$2^2 \cdot 3^2 = 4 \cdot 9 = 36$$

Now the written equation above can be expressed as:

$$c^2t^2 = t'^2 (c^2 - v^2)$$

That is correct and I will not use numbers again to check it, but if you are willing to do so, please do. In the next operation we can divide the equation by **c** squared:

$$c^2t^2 = t'^2 (c^2 - v^2) \qquad \text{divided by } c^2 \text{ is}$$

$$\frac{c^2 t^2}{c^2} = \frac{t'^2(c^2 - v^2)}{c^2}$$

$$t^2 = t'^2(1 - \frac{v^2}{c^2})$$

We can now remove square by applying squared root and in doing so, obtain the following equation:

$$t = t' \sqrt{1 - \frac{v^2}{c^2}}$$

The above equation shows the relation between time elapsed in one inertial reference frame **t** (the man standing on the ground) and the time **t'** elapsed for the man standing in the train as perceived by the man standing on the ground. This is basically the **Lorentz transformation** applied to time difference as a result of a different speed between our reference frame and another reference frame when we measure the time in our reference frame and the time in the other reference frame. The equation shows that the higher the speed of the other reference frame, the slower the time is for that reference frame as perceived by us.

We can see that in the situation where **v** reaches the speed of light, then **t'** is equal zero and this makes perfect sense.

We remember our girl on the train moving at the speed of light and looking in the mirror. Photons from her face move at the speed of light for her which is 300 000 km/sec. These photons move at the same speed for the man standing on the ground. As she herself moves at the speed of light (she is on a train moving at such a speed) then the only way the man on the ground can see her photons moving at that speed is if time has stopped for the photons leaving her face. In other words, if time for these photons is zero as well as the distance from the mirror and her face. In that case, the photons from her face move in zero time and pass zero distance for the man on the ground but are perceived as moving at the speed of light because her train is moving at that speed. Mathematically, if **v** is equal **c** we have:

If $v = c$ then

$$t = t' \sqrt{1 - \frac{c^2}{c^2}} = t' \cdot 0$$

c/c is 1 and 1-1 is 0

The Lorentz transform for length contraction is:

$$L = L' \sqrt{1 - \frac{u^2}{c^2}}$$

u = relative velocity between two reference frames.

Again, we have a situation where if relative velocity or difference in speed between two reference frames is equals c then l' is equals zero.

THE TWIN PARADOX

Alison works for NASA. She is the creation of my imagination in my book so I can do whatever I want with her. She attended this business meeting as her twin sister Caroline and herself were chosen to travel in space.

I forgot to mention that it is a story that takes place far away in the future when mankind has developed to the point that they have a spaceship travelling at the speed of 3/5 of the speed of light. People also construct space stations at different distances from the Earth. Somehow they build the space station, which is exactly 3 light years away from Earth.

Both twins are thirty years old and were preparing for this journey. However, at the time of departure Caroline became unwell so it was only Alison who joined the crew. She therefore left Earth at the speed of 3/5 of the speed of light and travelled 5 years until her ship reached the space station. She returned back to Earth at the same speed of 3/5 of the light speed and it took her again 5 years to reach Earth.

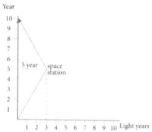

Picture 6.10

Caroline continued her life on Earth for 10 years. From her point of observation in her reference frame at rest, the time in the spaceship where Alison is passes slower than for Caroline on Earth. We can imagine that for this purpose the Earth is at rest in relation to the spaceship. Caroline is therefore not moving through space but is moving through time. Even if you stand still, time is passing by. On the space-time diagram she is moving only along a vertical line representing time, as she has a zero movement in space, on the horizontal line. Alison is moving at 3/5 of the speed of light according to Caroline's reference frame so she sees her sister, Alison, as aging more slowly than her. However, Alison sees Earth as moving away from her and perceives her ship as being at rest. From her reference frame, Caroline is the one who is aging more slowly. If Alison continues to travel until she dies then both sisters will perceive that it is the other one which is aging more slowly. As this will involve Alison travelling in one direction constantly until she dies, well passing the station at the distance of 3 light years, then the twins will never meet. There will not be the twin paradox, as they will never face each other.

Alison was, however, travelling in one direction for 5 years to the space station and was then instantly moving in the opposite direction for 5 years back to Earth.

The question is: which twin will be younger when they meet after 10 years?

Each of them thinks that the other one has aged more slowly. They cannot both be right.

The answer to this question is that Alison did not

travel in one direction all her journey. After 5 years of travelling, her ship changed direction to return to Earth. The changing of direction of velocity, even if a constant speed, means that acceleration is happening. Alison is in this scenario really moving according to the reference frame of Caroline.

Why does acceleration determine that the one who accelerates will be in the reference frame where time slows down (time dilatation)?

The answer to that question is easy. Imagine that two spacecrafts move at the same speed of 3000 km per hour. Both of them will be in the same reference frame at the same speed and will not move in relation to each other. Now if one of them accelerates then it will really move in relation to the other and will be therefore the one where time dilatation will be taking place and the crew of this ship will age slowly according to the ship that does not accelerate but remains travelling at the speed of 3000 km/hour.

We can now work out how old Alison will be when she returns to Earth. Caroline will age ten years and will be forty years old. Both of them were thirty when Alison left Earth speeding at 3/5 of light speed towards the space station. Her age upon return can be calculated by using the Lorentz transform, which is:

$$t = t' \sqrt{1 - \frac{v^2}{c^2}}$$

t is Caroline time while **t'** is Alison time.

v is her 3/5 of the light speed. So we can write fraction divided by fraction. In such mathematical calculations, it is the rule to multiply the outside number and put this product above the line. The product of multiplication of inner numbers comes under the line to make it only one fraction:

$$\frac{\dfrac{3}{5}}{\dfrac{5}{5}} = \frac{15}{25} = \frac{3^2}{5^2}$$

So:

$$t = t' \sqrt{1 - \frac{3^2}{5^2}}$$

Therefore:

$$t = t' \sqrt{\frac{25}{25} - \frac{9}{25}}$$

So:

$$t = t' \sqrt{\frac{16}{25}} = \frac{4}{5}$$

Or:

$$t = t' \ \frac{4}{5}$$

Therefore: 10=10 times 4/5 is 10=40/5, which is 10=8

While Caroline will have aged ten years and will be

forty years old, Alison for the same time will age only eight years and will be thirty-eight years old.

MINKOWSKI SPACE TIME

Herman Minkowski was a German mathematician who came upon the idea of combining three space dimensions and one time dimension into four-dimensional space-time called ' Minkowski space' or space-time which provided the mathematical foundations for Albert Einstein's special theory of relativity.

In relation to so-called non-Euclidian geometry, **Euclid** was a mathematician who lived in Alexandria, Egypt. He was born in 300 BC. He came up with five famous postulates, which are also known as axioms, which contributed significantly to our understanding of geometry related to flat geometry. In this geometry, the shortest distance between two points is a straight line. Based on this geometry, the sum of the angles inside a triangle is 180 degrees. Also based on Euclidian geometry, which is referred to as flat geometry, the distance between two points can be found by applying Pythagoras' theorem. If we draw lines through each point (A) and (B) so that these lines meet each other at the right angle then we can work out the value of the distance between A and B which will be the hypotenuse. Hypotenuse squared is equal to the sum of squared sides.

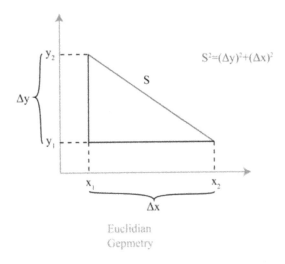

$$S^2=(\Delta y)^2+(\Delta x)^2$$

Euclidian
Gepmetry

Picture 6.11. Δis a Greek symbol used to mark the difference in value as the difference between X1 and X2 or Y1 and Y2 in the above illustration.

On a circle, however, or on a square, the shortest line between A and B is not a straght line but a curved line. As it is not a straight line, it is rather called *geodesic* which defines the shortest part between A and B. Also, the sum of the inner angles of the triangle is not equal to 180 degrees but is more. If we try to draw a triangle starting from the North Pole and try to make a triangle of equal sides taken from the North Pole to the equator, then we can easily see that the angle between each side is 90 degrees, making the sum of the inner angles to be 270 degrees.

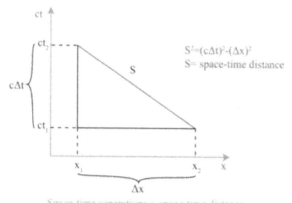

Space-time separations = space-time distance

- (Δx) Difference in space distance
- (cΔt) Difference in time distance
- S Space-time separation

Picture 6.12

To go back to Minkowski space-time, as he combined three dimensions of space and one dimension of time he got an equation where hypotenuse squared is equal to time squared minus space squared.

S is the symbol for space-time interval or separation in time and space. Their value depends on the speed or velocity at which a particular object or event is in relation to the observer. However, when calculated using the above equation, then the space-time interval does not change and is the same regardless of another reference frame or place from where it is observed. It is obviously for the particular reference frame or velocity the particular object is travelling. Because it is the same, it presents invariance obtained by using this calculation. In the next illustration is an example of space-time separation at velocity of the speed of light. As we can see, we have a time of 3 years and distance light passed for 3 years (3 light years). As **S** is equal time **t** minus space **x** we have **S** equals zero.

We have therefore space-time interval or space – time separation zero at the speed of light. We can just remind ourselves of the example of the girl who is in the train speeding at the speed of light at 300 000 km/second and looking at herself in the mirror. The man on the ground was able to see light (photons from her face) going at the speed of 300 000 km/second as well as the girl on the train because the space between the mirror and the girl was zero as well as the time passed for photons leaving her face to reach the mirror. This is exactly what we get by applying the above equation to calculate the space-time interval or separation for an inertial frame of reference which travels with the speed of light. In other words, the space-time interval or separation is zero.

The trajectory of a movement at a certain speed in the space-time diagram is called the *world line* to differentiate it from using the term trajectory. The reason for this is that the world line marks the movement of an object in a space-time diagram not only in three-dimensional space where the term trajectory is used. The space-time separation for a particular world line is the result of time (t) minus space (x). We have a positive result for all world lines referreing to an object travelling at a speed less than the speed of light. That refers to any world line, which is between time coordinates **ct** or **t** and 45 degrees of light world line. Such space-time interval is called **timelike** as time separation is greater than space separation.

If an object takes light like a world line travelling at the speed of light, then space-time separation is zero. This object has to be massless, as we know that relativistic mass is the sum of rest mass and the speed. If an object has a mass then the mass will increase with the speed, and with it, energy needed to increase the speed of the object. If we want to push the speed of the object to the speed of light then its mass will increase infinitely and with it the energy needed to push this object to the speed of light. That is why an object has to be massless to be able to travel at the speed of light. In other words, an object that has mass at rest will never reach the speed of light as it requires infinite energy due to relativistic mass increasing towards infinity as its speed increases towards light speed.

If the space-time separation is a negative result then world lines are based between light world line of 45 degrees and the horizontal axis **x** presenting space separation. Such world lines are called **spacelike** as space separation is greater than time separation.

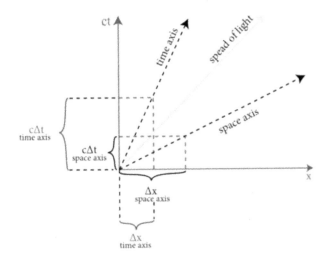

Picture 6.13

One important quality of constant and limited maximum speed in our universe, which is preserved for light speed, is that limitation of maximum speed presented by light speed secured the **causality principle** in the universe. This means that we can never encounter effect before its cause. Cause always has to precede the effect. Perhaps the simplest way to explain this is with the help of the illustration below:

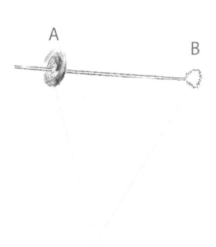

A

B

O

Picture 6.14

A supernova has exploded at point **A** and has produced a gamma ray burst which travels at the speed of light, reaching the planet at point **B**. The result is the planet's destruction.

The light from point **A** at the moment of the supernova explosion and the creation of the gamma burst reaches us at a particular time.

Light reflecting the destruction of the planet at point **B** will reach us at a particular time but will never be at the same time with the light from point **A** at the moment of explosion and the creation of the supernova and gamma rays. The reason is that light has a speed limit and it will take time to pass the distance from **A** (cause) to **B** to carry information which will result in the destruction of the planet (effect). The distance between **A** and **O** is less than the distance between **A** plus **B** plus **O**. We cannot have information about planet destruction before this information is delivered to point B from point A. All of this shows that cause is always seen first or chronologically it happens before effect. Therefore, the causality principle is secured by constant and limited maximum speed represented by the speed of light.

Simultaneity is another important event taking place in the universe, which can be perceived differently.

If we are referring to an event which is taking place at the same time and in the same place, then it will be seen as simultaneous regardless of the distance from which it is observed in space and time or in which reference frame is an observer. An example of this is a traffic accident when two cars collide, clapping with fingers or kissing. All these events take place at the same time and same place.

If we are talking of two traffic accidents happening at different places but at the same time then their simultaneity depends on the inertial reference frame an observer is in. Simultaneity is therefore relative. To understand this we should remind ourselves of the two main postulates of the special theory of relativity. In the first, it is stated that all laws of physics remain the same in every reference frame.

If I am travelling with the train I consider myself to be at rest. If I put myself in a time- space diagram I will move only along vertical axes representing time, as

I believe that I am at rest and not moving in a horizontal line or through space. If two balls fall from a luggage compartment at the top of my carriage, one at the front and one at the back of the compartment at the same time, then I will perceive two events happening simultaneously. This is, however, in relation to my space axis which is at the angle with the time axis of my time in my referential frame. The proper time is the time shown by the clock, which I carry with me in my referential frame. This has invariant property upon which everybody agrees.

The second postulate is that the speed of light is always the same regardless of the reference frame of the observer (whether the observer is at rest or is at constant velocity).

To explain the above better, I believe that the next few illustrations will be helpful.

First, let's imagine an observer is at rest. This is shown in the next diagram. There are also three events taking place, which are for this observer (being at rest) simultaneous.

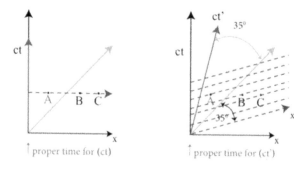

Picture 6.15

In the above diagram concentrate for a moment only on the left-hand diagram and disregard the right-hand one to avoid confusion. The left-hand diagram is a diagram of an observer at rest who moves through his time along vertical time axes. At one point he will observe a simultaneous event taking place at A, B and C. It is according to his space axis, which is at 90 degrees with his time axis. We should not forget that light speed is taking a diagonal line, which is equally 45 degrees from time and space axes.

Forget for the moment the right-hand side of the diagram representing an observer in movement at constant velocity and concentrate on the next diagram. In the next diagram I am trying to draw a diagram of an observer who is moving at constant velocity CT'. We should not forget that this is a diagram of an observer who is moving at constant speed, combined with a diagram of the observer who is at rest. We have here two diagrams fused together. One, of an observer at rest CT (time axis) – X (space axis), and the other, of an observer who is moving at constant speed CT' (time axis) X' (space axis). He carries his watch with him, which shows him a proper time for his frame of reference where he considers himself at rest. He therefore moves only along his time axis CT'. However his time axis is moving with constant velocity in relation to the diagram of the observer who is at rest. That is why CT' will be at an angle compared with CT. Now if we want to construct his space axis (X'), it should be at a right angle with his time axis (CT') as shown in the right-hand diagram. This, however, is not possible because light will be going at C'. Although this would be at 45 degrees for his frame CT' and X' which is the speed of light, it will not be from the reference frame of the man at rest in the diagram.

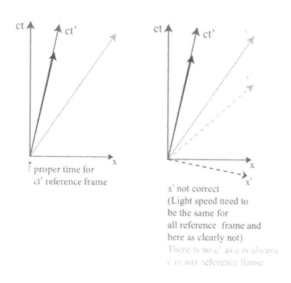

Picture 6.16

In the reference frame of the observer at rest, C' will be faster than the speed of light as it is less than 45 degrees with his X or space axes. The second postulate states that the speed of light is the same for all reference frames. We should notice that the speed of light forms the same angle with CT and X axis and that should be in all reference frames. We also need to have the speed of light at the same place in all reference frames. Therefore, the logic conclusion is that X' should be at the same angle with the line of the speed of light like it is CT' as in the next diagram.

Now the space axis of X' is at the angle with the space axes of an observer in a referential frame at rest. If an observer in constant velocity at CT' considers

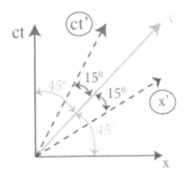

Picture 6.17

he is at rest taking his CT' as his proper time axis then only events happening in his space axis X' will be simultaneous for him (the balls falling from the compartment in his carriage at the same time in the front and at the back of his carriage). However, events which take place simultaneously for an observer at rest (A B and C) in the first diagram on the left will not be simultaneous for an observer at reference frame CT' as his X' axis will be at an angle with the X axis. Now concentrate on the right-hand side of the illustration and diagram of an observer who is moving and in the reference frame CT'. He will observe first C, then B, and finally event A. To make it easier, I'll repeat this illustration below. You can see that CT' is at 35 degrees with the light speed world line and the same angle of 35 degrees is between the light speed world line and X' axis.

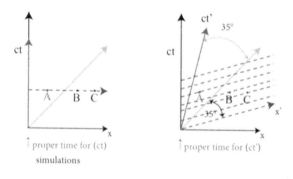

Picture 6.18

GENERAL RELATIVITY

I have made an effort to explain the special theory of relativity focusing on particular points which were confusing to me initially. These points prevented me having a better understanding of this concept when I read about it for the first time. I hope that I have managed to explain this in a way that makes sense. In doing this, I did not go into a very detailed explanation regarding the Lorentz transformation, for example. Nevertheless, I believe I have provided a decent amount of material there.

As far as the general theory of relativity is concerned, I will not really go into so much detail, not only because it is not necessary for the purpose of my book but because it is a very complicated topic which requires a high level of mathematical skill and knowledge that are well beyond my capability.

The essence of the general theory of relativity is that gravity is understood as the result of space-time geometry or precisely it is caused by acceleration due to the fact that space-time is curved.

Einstein finished his general theory of relativity in 1916. It is based on two postulates:

1. *Relativity principle*
   Local physics is governed by the theory of special relativity
2. *Equivalence principle*
   There is no way to make a difference locally between gravity and acceleration

Einstein's approach initially to solving a gravity problem was to apply a famous thought experiment. He thought about a man standing in a lift where the lift suddenly gets detached from its main cable. It is the situation where the man and lift suddenly start to accelerate equally towards the Earth. As both accelerate or speed up at the same rate, then both the lift and the man fall towards the bottom of the building at the same rate of speed, which is experienced as weightlessness. If the man had a ball in his hand and he decided to let go of the ball from his hand then the ball would remain in the air next to his hand after he let it go from his hand. The reason is that all of them – the man, the ball and the lift – are speeding up at the same rate towards Earth. It is the same principle which is used to experience weightlessness in an airplane used for this purpose. The airplane flies high in the atmosphere and then lets gravity take the plane down for a while. It is a free fall due to acceleration caused by gravity where all the objects in the plane fall at the same rate involving people. That is how weightlessness is experienced.

Einstein has imagined also the situation where a man is in a lift in the space accelerating at the same rate. With no window in this lift, the man will not be able to know whether he is in space or he is falling to Earth due to acceleration.

The next experiment demonstrated that light beams bend due to acceleration.

To make it easier to understand, let's try to imagine a lift which moves upwards with a constant velocity. *There is no acceleration, but constant velocity or an inertial reference frame.* If we shine a light at one moment from the left-hand side of this lift, then light will travel towards the other side of the lift and will reach the other side of the lift after a short time. Let's say the lift was travelling at a constant speed of 20km/hr. At the moment that light enters the space from the lift it was travelling upwards at 20 km/hr. The light will remain at this speed upwards. As both the light and the lift continue at the same speed upwards then light will reach the other side of the lift at the same height from where it left the left-hand side of the lift. In other words, the light beam will be parallel with the floor or ceiling of the lift. The man in the lift will see that light beam as straight light. I did not draw the man in the lift again in order to give a clear illustration, but you can just imagine the man standing in the lift.

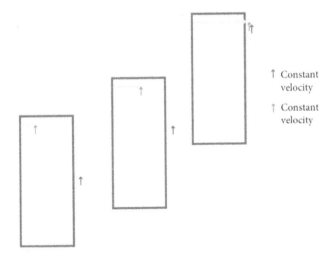

Picture 6.19

Let's now imagine that this lift moves upwards but is accelerating or its speed is changing.

We now have acceleration with changing velocity or

a non-inertial reference frame. We now shine light again from the left-hand side of the lift when the lift was at the speed of 20 km/hr. The light will continue to move upwards at the speed of 20 km/hr, while it is moving horizontally towards the right-hand side of the lift. However, during the time the light or beam of this light is travelling through the space from the left at the constant speed of 20 km/hr upwards, the lift is already moving upwards at the speed of 60km/hr as it is constantly accelerating upwards. So by the time the light beam passes half of the space on its way to the other side of the lift, the lift has moved significantly upwards. It will have moved even more by the time the beam of light reaches the other side of the lift. So although the light travels in a straight line from the left-hand side to the right-hand side side of the lift, it enters the lift on the left-hand side at the top of this side and reaches the other side of the lift close to the point where the right-hand side meets the floor of the lift. For the man standing in the lift, the beam of this light has bent due to acceleration, which is the result of gravity. I again tried to demonstrate it with an illustration with no man inside the lift for the sake of clarity of the illustration. Here, however, I did not show all the light beam but just one of the photons which goes one after another making a straight line of a light beam: similarly to drops of water that go so close to each other at such a speed that altogether they give the impression of a water jet. In the following illustration it can be seen how the photon of that light (all of these photons make this particular light beam) leaves the lift on the left-hand side at the top of the wall and hits the other side of the lift wall at the bottom. This happens because now the lift does not go up at a constant speed but accelerates.

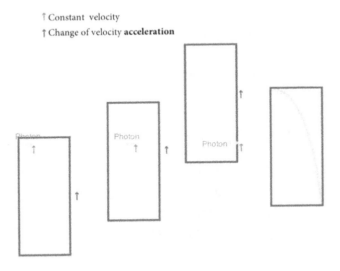

Picture 6.20

According to Newton's law of gravitation, the force of gravitation is equal to the product of **G** gravitational constant, **M** mass of one object and **m** mass of another object divided by **r²** radius on squared:

$$F = \frac{G\,M\,m}{r^2}$$

This equation cannot be applied to light, as light is massless. Einstein therefore rejected mass as the reason for gravitational attraction or force. Instead, he believed that space-time is curved which makes light bend. Light continues to go in a straight line but as it comes across curved space-time it takes the shortest part or geodesic.

If we look again at the space-time diagram which shows movement with constant velocity of the object moving in an inertial reference frame then we can see that every world line is straight due to constant velocity (left-hand diagram on illustration below). If, however, we have acceleration then the world line becomes curved (right-hand diagram in the illustration). The space is curved due to the massive object, which is manifested with acceleration. The movement which is non inertial (acceleration or changing velocity) is a real movement in relation to movement with constant velocity. It causes dilatation of the time and contraction of the space as manifestation of curved space.

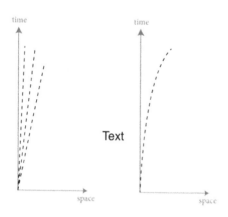

Picture 6.21

Einstein worked for around ten years to finalise his general theory of relativity. In this process he had the support of many great minds in physics and mathematics at that time. **Handrik Lorentz**, Dutch physicist, was one of them. He calculated the famous transformation inspired by Albert Michelson and Edward W. Morley's discovery that the speed of light is constant and that there is no ether. It was, however, Albert Einstein who put things together and created the general theory of relativity, which has been tested and proved over and over.

As we all know, the satellite navigation system that we all use in our cars, is controlled by satellites located in orbit around Earth. As they are at a distance from the Earth, their acceleration is slightly weaker than on Earth. So objects on Earth are moving slightly faster, which means that their clocks are slightly slower than the ones in satellites. This needs to be taken into account and calculations need to be made for clocks in satellites. If this is not done, then very quickly their guide will become out of precision and will not be useful. The difference in time is incredibly small but would quickly become significant for devices not to be able to function properly.

Einstein became the most celebrated scientist in the twentieth century. He was seen on many occasions and in the company of many celebrities at that time. On one such occasion he was invited to the premiere of the Chaplin film *City Lights* where he appeared together with Charlie Chaplin. When they were mobbed by thousands, Chaplin remarked, "The people applaud me because everybody understands me, and they applaud you because no one understands you." Einstein asked Chaplin, "What does it all mean?" Chaplin replied, "Nothing."

SUMMARY

The universe is composed of everything that exists. It was created around 13.7 billion years ago in the Big Bang explosion, details of which I will focus on in the next part of the book.

The main components of the universe such as time, space, matter, dark matter, fundamental forces, energy and dark energy were all created in the Big Bang explosion. Before that, nothing existed.

Scientists have calculated that matter occupies only 4% of the universe while dark matter makes up 23% and the remaining 73% is made of dark energy.

Ordinary matter is composed of elementary particles. There are altogether only twelve elementary particles from which all matter of the universe is made up. There are six quarks and six leptons with the difference between them being such that quarks are affected by strong nuclear forces while leptons are not.

The elementary particles have one important property referring to their spin of rotation around their own axes. According to this property, all particles are divided into those which have ½ of a spin called fermions, and those with a whole spin called bosons. Fermions are subject to the Pauli principle of exclusion. That roughly means that two fermions cannot be in the same place at the same time. The fermion's job is to create all matter while the boson's job is to establish communications between particles in each of four fundamental interactions. Bosons are not subject to the Pauli principle of exclusion, which means that an unlimited amount of bosons can be in the same place at the same time.

With the creation of matter, all four fundamental interactions between particles came to light and were responsible for the arrangement of matter at microscopic and macroscopic levels.

At a microscopic level, strong nuclear interactions created strong, tight connections between three quarks forming protons and neutrons. Weak nuclear interactions enable the transformation of one proton to one neutron, which sets part for the formation of a helium atom, for example. The next electromagnetic fundamental interaction helps the creation of stable atoms consisting of protons and neutrons in an atomic nucleus and electrons orbiting around the nucleus at a distance. An atom with a different number of protons inside had a particular specific property of so-called elements of which this atom was the smallest part. These atoms belonging to different elements were able to combine among them, creating compounds with a different property. This was possible thanks to electromagnetic fundamental interactions. We can say that strong and weak nuclear interactions together with electromagnetic fundamental interactions help the arrangement of matter at a microscopic level.

The arrangement of matter at macroscopic level or at the level of the universe in the form of stars, planets and galaxies was possible thanks to the fourth fundamental interaction – gravitation.

Sir Isaac Newton formulated gravitation as a force, which depends on the product of gravitational constant, and two masses, which attract each other, divided by radius of their separation squared. It was during the time when it was believed that time and space had absolute value and as such had invariant property. With the discovery of invariant property of the speed of light being at a constant speed of around 300 000 km/second, regardless of the reference frame from where it is observed, Einstein developed his special and general theory of relativity. The last one, on which he had been working for almost 10 years following the special theory of relativity, explained gravitational force from a completely different angle. It basically states that gravitational force is the result of curved space-time fabric where the quantity of the mass dictates how much space- time will be curved and the curvature of space-time will determine the movement of another mass that approaches this curved space-time fabric. Time and space are not invariant but are flexible and changeable which is experienced as time dilatation and space contraction when those two properties are observed in a non-inertial reference frame (which is acceleration or gravitational force) in relation to an inertial reference frame (which is state at rest or movement at constant velocity).

Therefore, it could be said that three fundamental

interactions shape or arrange matter at a microscopic level while a fourth fundamental interaction – gravitation – shapes matter at a macroscopic level as well as time-space fabric itself.

Interactions among particles and the movement of matter within time-space are possible thanks to energy, which can be defined as the capacity to do work from the aspect of classic physics. Matter itself has energy, which is locked inside matter and is expressed by Einstein's famous equation $E = mc^2$. The equation represents formula, which helps to calculate the amount of energy, which mass contains when it is at rest; in other words, when there is no movement of the mass or added kinetic energy. It is also an expression of mass-energy equivalence where one can be transformed into another and vice versa. The previous knowledge of conservation of matter and energy as separate entities has to be defined more precisely as conservation of mass- energy within the closed system. It means that the amount of energy-mass created at the beginning of the universe has remained constant throughout time and space of the created universe although it is possible that one transfers to another to some extent and vice versa, but the total amount remains the same. These conservation laws are examples of invariance in modern physics. There are plenty of invariances such as the constant speed of light and the Planck constant, for example. There is also one of many philosophical approaches which tries to explain why the universe has developed the way it has. It is called the anthropic principle, which goes from the presumption that the universe has developed in this way with such particular forms of invariances or constants to allow life form to develop within such a universe.

Alongside energy, which can be described and formulated from the aspect of classical physics and the aspect of Einstein's special theory of relativity which referred to ordinary energy within the universe, there is also energy, which is the opposite of gravity and tends to expand the universe. It is named dark energy.

Before this energy was discovered, we discovered another form of matter. This matter cannot be seen but was detected with the help of gravitational lensing and orbital velocity. With the help of gravitational lensing, it was discovered that dark matter is arranged like a spider's web throughout the universe and serves as scaffolding or support for building galaxies and clusters of galaxies of which ordinary matter is made.

As far as dark energy is concerned, it has been discovered that the universe has accelerated in its expansion over the last 5 billion years. This was detected thanks to distances between galaxies which were measured with the help of light emitted from type 1a supernovas which come from different galaxies (standard candles) and with the help of baryon acoustic oscillations (standard rulers).

From this recent discovery it can be argued that the future of our universe faces the so-called big reap where the universe will finally disappear, expending infinitely. Also, entropy where useful energy is transferred into less useful energy, all the time, will bring the universe to the point where the whole existing energy will be in a form of useless energy, bringing the universe to an end. In any case, the future looks quite bleak.

I myself now feel depressed as I am coming to the end of the first part of the book, but I am happy that I will start from the beginning of everything in the next part of the book.

Bibliography

Books

Foundations of modern cosmology, Hawley, John Frederic Holcomb, Katherine A.

The rough guide to the universe, John Scalzi

Chemistry essentials for dummies Moore, John T.
Atkins' physical chemistry, Atkins, P. W. (Peter William)
Higgs the invention and discovery of the 'God Particle', Baggott, J. E.

The bigger bang, Lidsey James E.

Higgs DiscoveryThe Power of Empty Space, Randall, Lisa

Antimatter the ultimate mirror, Fraser Gordon

Physics II for dummies, Holzner Steven

The discovery of subatomic particles, Weinberg, Steven

Magnetism a very short introduction, Blundell, Stephen

On space and time, Connes, Alain Majid, Shahn

North Pole, South Pole, Turner Gillian

The Space Book, Jim Bell

Dark Matter, Dark Energy, Dark Gravity, Stephen Perrenod

Way does E=mc2?, Brian Cox, Jeff Forshaw

The First Three Minutes, Weinberg, Steven

Big Bang, Simon Singh

Astronomy A beginner's guide to the sky at night, Paul Sutherland

History of the Universe, Wyken Seagrave

Binding Energy, Fission and the Strong, Nuclear Force

Online Materials

http://hyperphysics.phy-astr.gsu.edu/hbase/amom.html

http://hyperphysics.phy-astr.gsu.edu/hbase/torq.html#torq

http://www.youtube.com/watch?v=s_R8d3isJDA

http://www.britannica.com/EBchecked/topic/181349/
Albert-Einstein/256585/World-renown-and-Nobel-Prize

http://abyss.uoregon.edu/~jscosmo/lectures/lec05.html

http://www.youtube.comwatch?v=UkLkiXiOCWU

http://www.bing.com/

search?q=proton+size&FORM=MSNSHL

http://map.gsfc.nasa.gov/universe/uni_shape.html

http://abyss.uoregon.edu/~js/cosmo/lectures/lec05.html

http://hyperphysics.phy-astr.gsu.edu/hbase/amom.html

http://hyperphysics.phy-astr.gsu.edu/hbase/torq.html#torq

http://www.youtube.com/watch?v=s_R8d3isJDA

http://www.britannica.com/EBchecked/topic/181349/Albert-Einstein/256585/World-renown-and-Nobel-Prize

http://www.britannica.com/EBchecked/topic/348136 Lorentz-

FitzGerald-contraction
http://www.britannica.com/EBchecked/topic/348120/hen-

Hendrik-Antoon-Lorentz
http://www.britannica.com/EBchecked/topic/384284/ Hermann-Minkowski

http://www.britannica.com/EBchecked/topic/557482/ spacetime# ref206206

http://www.britannica.com/EBchecked/topic/194880 Euclid

Special Relativity - A Level Physics
http://www.youtube.com/watch?v=6Tts3gxs_cM

Spacetime and the Twins Paradox
http://www.youtube.com/watch?v=KdjQkuGTBMo

Einstein Field Equations - for beginners!
http://www.youtube.com/watch?v=foRPKAKZWx8

Hubble, Big Bang and The Age of the Universe - Part 1
http://www.youtube.com watch?v=S3ZWKWw5mPA&list= PL72DD6CBED89C0A06

Prof. Brian Cox - A Night with the Stars BBC Full Lecture
http://www.youtube.com/watch?v=5TQ28aA9gGo